Lecture Notes in Mathematics

An informal series of special lectures, seminars and reports on mathematical topics

Edited by A. Dold, Heidelberg and B. Eckmann, Zürich

21

A. Borel · S. Chowla · C. S. Herz
K. Iwasawa · J-P. Serre

Seminar on Complex Multiplication

Seminar held at the Institute for Advanced Study,
Princeton, N.J., 1957-58

1966

Springer-Verlag · Berlin · Heidelberg · New York

Library of Congress Catalog Card Number 66–27636. Title No. 7341

TABLE OF CONTENTS

I STATEMENT OF RESULTS

(J-P. Serre, Oct. 16, 1957)

§1. The notion of complex multiplication.

Let X be an elliptic curve. As a complex Lie group, it is the quotient of the complex plane $\mathbb{C}$ by a lattice Γ, spanned by two periods ω_1, ω_2, and since X is isomorphic to the curve defined by the periods $z\omega_1$, $z\omega_2$ for any non zero $z \in C$ we may assume Γ to be spanned by 1 and τ, where τ has a positive imaginary part.

An endomorphism of X may be identified with an endomorphism of its universal covering $\mathbb{C}$ mapping Γ into itself; it is therefore the multiplication by a complex number z such that $z, z\tau \in \Gamma$. The endomorphisms of X form a ring A(X), which always contains the integers $\mathbb{Z}$, (the "trivial endomorphisms"). The other ones (if any) are given by complex numbers and are called complex multiplications. If $A(X) \neq \mathbb{Z}$, the curve X is said to admit complex multiplications.

"In general", X has no complex multiplication. In fact, assume that z defines a non trivial endomorphism of X. Then

$$z = a + b\tau \ , \ z\tau = c + d\tau \ , \quad (a,b,c,d \text{ integers, } b \neq 0) \ ,$$

whence

$$a\tau + b\tau^2 = c + d\tau$$

and τ must belong to an imaginary quadratic field, say K; moreover z belongs to the ring of integers $\underline{o}(K)$ of K since it is in K and defines an endomorphism of a $\mathbb{Z}$-module of finite rank, namely Γ. Therefore, A(X) is an order of K, (subring of $\underline{o}(K)$ containing $\mathbb{Z}$ and which has rank 2 as a $\mathbb{Z}$-module);

one gets in this way all orders of all quadratic imaginary fields (if R is such an order, take for X a curve with lattice of periods R; since $1 \in R$, $zR \subset R$ if and only if $z \in R$, whence $A(X) = R$).

Assume that $A(X) = \underline{o}(K)$, and that $\Gamma \subset K$. Then Γ is an ideal of K, and conversely any ideal of K gives rise to a curve X such that $A(X) = \underline{o}(K)$. Two such curves are isomorphic if and only if the corresponding ideals are homothetic, i.e. belong to the same ideal class.

Let j be the modular function. For the curve with normal equation

$$y^2 = 4.x^3 - g_2.x - g_3$$

its value is

$$j = 2^6.3^3.g_2^3/\Delta \quad , \quad (\Delta = g_2^3 - 27.g_3^2) \ .$$

Two elliptic curves are isomorphic over an algebraically closed field if and only if their modular invariants are equal. By the above, j defines a function on the ideal classes $\underline{k}_1, \ldots, \underline{k}_h$ of K; the numbers $j(\underline{k}_i)$ are "singular values" of j, and are called the class invariants of K; they are pairwise different, and have proved to be of fundamental importance in the study of the abelian extensions of K, to which we now turn.

§2. Unramified abelian extensions of an imaginary quadratic field.

It is a classical result of Kronecker that every abelian extension (i.e. normal extension with commutative Galois group) of the field $\mathbb{Q}$ of rational numbers is contained in a field of roots of unity. Thus, so to say, certain values of the exponential function generate the maximal abelian extension of $\mathbb{Q}$. Such an "explicit" construction is also possible for an imaginary quadratic field. One has to use the class invariants and also the values of a certain function, related to the Weierstrass p-function (see §4). This theory is essentially what is called "complex multiplication". We shall first deal with unramified extensions. The results pertaining to this case may be embodied in the following three theorems, where K is an imaginary quadratic field, $\underline{k}_i$, $(1 \leq i \leq h)$, its ideal classes.

THEOREM I. *The class invariants* $j(\underline{k}_i)$ *are algebraic integers.*

[Let us remark in passing that there is a sort of converse to Theorem I. Namely, C. L. Siegel (Transcendental numbers, Annals of Math. Studies 16, Princeton, 1949, pp. 98-99) has deduced from certain results of Schneider that if z is an algebraic number in the upper half-plane not belonging to a quadratic imaginary field, then j(z) is transcendental.]

THEOREM II. $K(j(\underline{k}_i))$ *is independent of* i, $(1 \leq i \leq h)$, *and is the maximal unramified abelian extension of* K.

(Unramified means that every prime ideal of K decomposes in a product of distinct prime ideals with exponents 1.)

By class field theory, it is known that the maximal unramified abelian extension of K (Hilbert's "absolute class field") exists and that its Galois group G_K is canonically isomorphic to the group C_K of ideal classes; the next theorem describes how it operates on the $j(\underline{k}_i)$.

THEOREM III. *Let* $k \in C_K$ *and let* $\sigma_{\underline{k}} \in G_K$ *be its image by the isomorphism of class field theory. Then*

$$\sigma_{\underline{k}}(j(\underline{k}_i)) = j(\underline{k}^{-1}.\underline{k}_i) .$$

For the proofs of the above theorems, we shall follow the method of the first part of Hasse's paper quoted in §5, which presupposes class-field theory. More specifically, Hasse establishes by function theoretical arguments Theorem I and the congruence

(1) $$j(\underline{p}^{-1}.\underline{k}) \equiv j(\underline{k})^{N\underline{p}} \bmod \underline{p}$$

($\underline{p}$ prime ideal of K, $\underline{k}$ ideal class, N$\underline{p}$ absolute norm of $\underline{p}$), for almost all (i.e. all but a finite number of) prime ideals of K having absolute degree one (i.e. N$\underline{p}$ = p with p prime) with respect to $\mathbb{Q}$. The validity of (1) for all $\underline{p}$ and Theorems II, III will then follow by class field theory.

§3. Algebraic interpretation.

For any complex number j, there is an elliptic curve X defined over $\mathbb{Q}(j)$, with invariant j; it may for instance be given by the equation

$$y^2 = 4x^3 - h(x + 1) , \qquad (h = 27j . (j - 2^6.3^3)^{-1}) .$$

If in particular $j = j(\underline{k})$, with K, $\underline{k}$ as before, then K(j) is the smallest field of definition for all elements of A(X). From this, it follows easily that the numbers $j(\underline{k}_i)$ are algebraic and that $L = K(j(\underline{k}_i))$ is independent from i.

Moreover, given a prime ideal $\underline{p}$ of K, there is an algebraic procedure to obtain from X a curve, call it $X_{\underline{p}}$, with invariant $j(\underline{p}^{-1}.\underline{k})$, also defined over K(j).

Deuring's method relies on the notion of reduction of X modulo a prime ideal $\underline{q}$ of L. This means that we consider the curve with the same equation as X, but the coefficients being reduced mod $\underline{q}$. It may be shown that for almost all $\underline{q}$, this is possible and leads to an elliptic curve $\bar{X}$ which does not depend on the particular equation chosen to define X. The curve $\bar{X}$ is then defined over a field of characteristic p, where p is the prime number contained in $\underline{q}$.

To any algebraic variety V defined over a field of characteristic p ≠ 0, one can associate its "p-th power" V^p. Roughly speaking, V^p is obtained by raising to the p-th power all coordinates of the points in any affine model of V. In particular we can consider $\bar{X}^q$ for any $q = p^s$ (s positive integer).

With these notations, Deuring proves by a simple argument that the equality

$$\bar{X}^{N\underline{p}} = \overline{(X_{\underline{p}})} \ , \ (\underline{p} = K \cap \underline{q}) \ , \tag{3}$$

holds for almost all $\underline{p}$ of degree 1, whence the congruence (1).

So far, it has not been possible to prove along these lines that the class invariants are integral numbers. Deuring's proof for this fact is the analogue with formal power series of the classical argument.

§4. Ramified extensions.

Let as before $K = Q(\sqrt{-d})$, (d positive square free integer), be a quadratic imaginary field, and X an elliptic curve with invariant $j = j(\underline{k})$, where $\underline{k}$ is an ideal class of K. Its group of automorphisms Aut X is the group of units of A(X), i.e. of $\underline{o}(K)$. This group is cyclic of order 2 (resp. 4, resp. 6) when $d \neq 1, 3$ (resp. $d = 1$, resp. $d = 3$).

The quotient X' of X by Aut X is an algebraic curve (of genus zero), defined over K(j).

Since X is a group variety, defined over K(j), its points of order n will be algebraic and so will be their images in X'. We can now state the main theorem of complex multiplication for ramified extensions:

THEOREM IV. *Let K be an imaginary quadratic field, j a class invariant of K, and X an elliptic curve with modular invariant j defined over L = K(j). The maximal abelian extension of K can then be obtained by adjoining to L the coordinates in X' = X/Aut X of all points of order n of X (n = 1, 2, ...).*

In order to express Theorem IV analytically, we introduce the function

$$\tau(u; \omega_1, \omega_2) = (-1)^{e/2} p^{e/2}(u; \omega_1, \omega_2) g^{(e)}(\omega_1, \omega_2)$$

where e = order Aut X, $p(u; \omega_1, \omega_2)$ is the Weierstrass function,

$$g^{(2)} = 2^7.3^5\ g_2.g_3/\Delta$$

$$g^{(3)} = 2^8.3^4\ g_2^2/\Delta$$

$$g^{(6)} = 2^9.3^6.g_3/\Delta\quad .$$

We have then:

THEOREM IV'. <u>Let</u> K <u>be an imaginary quadratic field</u>, j <u>a class invariant of</u> K, ω_1, ω_2 <u>a basis for some ideal of</u> K. <u>Then the maximal abelian extension of</u> K <u>may be obtained by adjoining to</u> L = K(j) <u>the numbers</u> $\tau(\frac{a\omega_1+b\omega_2}{n}; \omega_1, \omega_2)$, $(a,b,n\in\mathbb{Z},\ n > 0)$.

The maximal abelian extension of K can also be obtained by adjoining to K the roots of unity, the values j(z) of the modular function for <u>all</u> $z \in K^*$ having positive imaginary part, and square roots of elements in the field thus obtained. For this and its relation to the so-called Kronecker Jugendtraum, see Hasse's Klassenkörper Bericht, Jahr. Ber. D.M.V. 35, 1-55 (1926), §10.

§5. Bibliographical notes.

We content ourselves with some brief indications, without making any attempt towards completeness. As to the 19-th century literature, we just quote:

KRONECKER, Complete Works, Vol. IV, §XI, No. 14.

The first two systematic and detailed accounts are to be found in:

H. WEBER, "Algebra", Band III, 1908,

R. FUETER, "Vorlesungen über die singulären Moduls und die komplexe multiplication der elliptischen Funktionen", I (1924), II (1927), Teubner.

Weber's book contains most of the essential results. However, although Weber already introduces and obtains several properties of the function τ, both he and Fueter have to use other, more complicated, functions to generate the maximal abelian extension. That this could be performed by means of τ only was first shown by Hasse:

H. HASSE, "Neue Begründung der komplexen Multiplikation", Teil I, Crelle Journal 157, 115-139 (1927), Teil II, ibid. 165, 64-88 (1931).

Teil II and the above mentioned books follow the analytical method. Hasse's Teil I combines analysis and class field theory.

The purely algebraic approach was initiated and carried out by M. Deuring. See notably:

M. DEURING, "Algebraische Begründung der komplexen Multiplikation", Abh. Math. Sem. Hamburg, 16, 32-47 (1947).

M. DEURING, "Die Struktur der elliptischen Funktionenkörper und die Klassenkörper der imaginär-quadratischen Körper", Math. Annalen 124, 393-426 (1952).

It was clear from the outset that a main obstacle to a generalization of Deuring's methods to higher dimensional abelian varieties was the lack of a good theory for reduction mod p in algebraic geometry. This was recently supplied by Shimura (Amer. Jour. Math. 77, 134-176 (1955)), and was applied to higher dimensional extensions of complex multiplication by Shimura, Taniyama, Weil. See:

G. SHIMURA, "On complex multiplications", Tokyo Symposium on algebraic number theory (1955), 23-30.

Y. TANIYAMA, "Jacobian varieties and number fields", ibid., 31-45.

A. WEIL, "On the theory of complex multiplication", ibid., 9-22.

The two-dimensional case had been considered long ago by Hecke (following a suggestion of Hilbert), using analytical methods. See:

E. HECKE, "Höhere Modulfunktionen und ihre Anwendung auf die Zahlentheorie, Math. Ann., 71, 1-37 (1912).

E. HECKE, "Über die Konstruktion relative Abelscher Zahlkörper durch Modulfunktionen von zwei Variabeln", Math. Ann., 74, 465-510 (1913).

[(Added in 1965). The Shimura-Taniyama-Weil theory has been published:

G. SHIMURA and Y. TANIYAMA, "Complex multiplication of abelian varieties and its applications to number theory", Publ. Math. Soc. Japan, 6, (1961).

A systematic exposition of the analytic method is given in:

M. DEURING, "Die Klassenkörper der komplexen Multiplikation", Enz. Math. Wiss., Band I-2, Heft 10, Teil II.

For further results, see:

K. RAMACHANDRA, "Some applications of Kronecker's limit formulas", Annals of Maths., 80, 104-148 (1964).]

II MODULAR FORMS

(J-P. Serre, Oct. 23 and 30, 1957)

§1. The modular group.

Let $E \subset \underline{\underline{C}}$ be the upper half plane $I(w) > 0$; the modular group G is the group of automorphisms of E of the form:

$$w \longrightarrow \frac{aw + b}{cw + d} \qquad (a,b,c,d \in \underline{\underline{Z}},\ ad - bc = 1).$$

This group is the factor group of $SL(2,\underline{\underline{Z}})$ by its center $\{\pm 1\}$.

Let now Γ be a lattice in $\underline{\underline{C}}$; we can choose two generators w_1, w_2 of Γ such that $w_1 \wedge w_2 < 0$; they are determined up to a transformation of $SL(2,\underline{\underline{Z}})$. If we then put $w = w_1/w_2$, we have $w \in E$, and the orbit of w under G does not depend on the choice of w_1, w_2. Two lattices Γ and Γ' correspond to the same orbit if and only if they are homothetic, i.e. if the elliptic curves $\underline{\underline{C}}/\Gamma$ and $\underline{\underline{C}}/\Gamma'$ are isomorphic. Conversely, every orbit Gw corresponds to some lattice (for instance, $\Gamma = \underline{\underline{Z}} + w\underline{\underline{Z}}$). Hence:

PROPOSITION 1. The isomorphism classes of elliptic curves are in one-to-one correspondence with the orbits of G in E.

We denote by X the set E/G of the orbits of G in E. Our first task is to find a well-behaved set of representatives for X in E.

§2. Fundamental domain for the modular group.

Let T be the translation $w \longrightarrow w + 1$, and S the "inversion-symmetry" $w \longrightarrow -\frac{1}{w}$; one has $T, S \in G$ and $S^2 = 1$.

PROPOSITION 2. <u>Let</u> $D = \left\{ w \mid w \in E,\ |Rw| \le \tfrac{1}{2},\ |w| \ge 1 \right\}$. <u>Any</u> <u>orbit of</u> G <u>meets</u> D. Moreover, <u>two distinct points</u> w <u>and</u> w' <u>of</u> D <u>are equivalent under</u> G <u>if and only if either</u> $w' = T^{\pm 1} w$, $Rw' = \pm \tfrac{1}{2}$ <u>or</u> $w' = Sw$, $|w| = 1$.

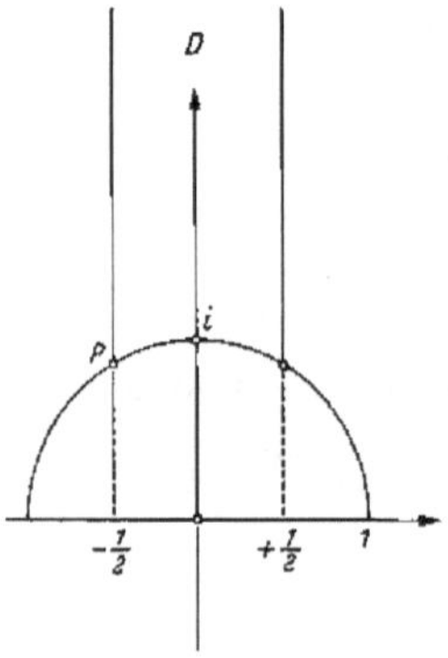

The set D is called a "fundamental domain" of the modular group G.

PROPOSITION 3. <u>The group</u> G <u>is generated by</u> S <u>and</u> T.

PROPOSITION 4. <u>The stability group of</u> $\rho = e^{2\pi i/3}$ (<u>resp. of</u> i) <u>is of order</u> 3 (<u>resp. of order</u> 2), <u>and is</u> <u>generated by</u> Q = ST (<u>resp. by</u> S). <u>Conversely, any point</u> <u>of</u> E <u>with a non trivial stability group is equivalent</u> <u>under</u> G <u>to either</u> ρ <u>or</u> i.

These three propositions will be proved at the same time. If $w' = Aw = \dfrac{aw + b}{cw + d}$, a simple computation shows that:

$$(1) \quad I(w') = I(w)/|cw + d|^2.$$

For a fixed $w \in E$, the set of all $cw + d$ ($c, d \in \underline{Z}$) is discrete; hence, the $I(w')$ have no non-zero accumulation point. Let then G' be the subgroup of G generated by S and T, and let G'w be any orbit of G' in E. By the above, we can assume that $I(w)$ is maximum on G'w; formula (1), applied with A = S, gives $|w| \ge 1$. On the other hand, $|R(T^n w)| \le \tfrac{1}{2}$ for a suitable $n \in \underline{Z}$; since T does not change imaginary parts, $I(T^n w) = I(w)$, and therefore $T^n w \in D$. Thus, <u>every orbit of</u> G' <u>meets</u> D.

Let now w and $w' = Aw$ ($A \in G$, $A \neq 1$) be points of D, with $I(w') \ge I(w)$. Formula (1) then yields $|cw + d| \le 1$, and, since we can assume $c \ge 0$, it follows that $c = 0, 1$. If $c = 0$, then $a = d = \pm 1$, $w' = w + b$, whence $b = \pm 1$ and $Rw = \mp \tfrac{1}{2}$. If $c = 1$, then we must have $|w| = 1$, $d = 0$, unless w is equal to ρ or $\rho + 1$, in which case $d = 0,1$ or $d = 0,-1$. The formula $d = 0$ gives $w' = a - \frac{1}{w} = T^a Sw$. Since $Sw \in D$, the first part of the discussion gives $a = 0$, except when $R(Sw) = \pm \frac{1}{2}$, i.e. $w = \rho$ or $\rho + 1$; when $w = \rho$, one can take $a = -1$, and, similarly for $w = \rho + 1$. Finally, the formulas $w = \rho$, $d = 1$ imply $w' = a - 1/(\rho + 1) = a + \rho$, whence $a = 0,1$.

Propositions 2 and 4 follow readily from this discussion.

Let now $A \in G$, and let choose a point w in the interior of D; the orbit of Aw under G' meets D. Hence, there exists $B \in G'$ such that $BAw \in D$. By propositions 2 and 4, we have $BA = 1$, hence $A \in G'$, and proposition 3 is proved.

Remarks.

1) It is possible to prove (for instance, by topological arguments) that the relations between S and T are generated by $(ST)^3 = 1$. Hence, the modular group G is isomorphic to the free product of a cyclic group of order 2 (corresponding to S) and a cyclic group of order 3 (corresponding to ST).

2) Let $Q(x,y) = Ax^2 + Bxy + Cy^2$ be a positive definite binary quadratic form, with real coefficients. Such a form can be written:

$$Q(x,y) = |xw_1 + yw_2|^2,$$

for a suitable choice of w_1, w_2. Applying proposition 2 to $w = w_1/w_2$, one then obtains the existence of a form $Q' = A'x^2 + B'xy + C'y^2$ which is equivalent to Q under the group $SL(2,\mathbb{Z})$, and verifies the inequalities:

$$A' \geq C' \geq |B'|$$

[This "reduction" process was already given in Gauss's Disquisitiones arithmeticae.]

§3.   Analytic structure and compactification of $X = E/G$.

We will first prove that the group G is "discontinuous" in E:

PROPOSITION 5.   Let K be a compact subset of E.   Then:

(i)   There exists a number N such that $I(Aw) < N$ for every $A \in G$ and $w \in K$.

(ii)   For every compact subset K' of E, the set of $A \in G$ such that $AK \cap K' \neq \emptyset$ is finite. [i.e. lim. $AK = \infty$].

Statement (i) follows from formula (1) and the fact that Inf. $|cw + d| > 0$, when w runs through K, and c,d run through all pairs of relatively prime integers.

Statement (ii) does in fact hold for every discrete subgroup of $SL(2,\underline{R})/\{\pm 1\}$.   In the case of the modular group G, it can be checked in the following way:

Formula (1) shows that the number of pairs (c,d) associated with transformations $A \in G$ such that $AK \cap K' \neq \emptyset$ is finite.   But two elements A_1, $A_2 \in G$ have the same (c,d) if and only if $A_1 = T^n A_2$ with some $n \in \underline{Z}$. Since $T^n A_2 K$ tends to ∞ with n, one has $T^n A_2 K \cap K' \neq \emptyset$ only for a finite number of values of n, and (ii) follows immediately.

COROLLARY 1.   The factor space $X = E/G$ is Hausdorff (and hence locally compact).

This is a formal consequence of (ii):

Let w and w' in E be inequivalent under G.   Since $Aw' \longrightarrow \infty$, there exists a compact neighbourhood U of w such that $Aw' \notin U$ for any $A \in G$.   Let V be a compact neighbourhood of w' and $N \subset G$ be the set of all A such that $AV \cap U \neq \emptyset$;   by (ii), the set N is finite.   For any $A \in N$, let W_A be a neighbourhood of Aw' which does not meet U, and put $U' = V \cap \bigcap_{A \in N} A^{-1} W_A$;   the set U' is a neighbourhood of w'.   If $A \in N$, one has $AU' \subset W_A$ hence $AU' \cap U = \emptyset$; if $A \notin N$, one has $AU' \subset AV$ hence again $AU' \cap U = \emptyset$. This means that GU and GU' are disjoint saturated neighbourhoods of the orbits Gw and Gw', q.e.d.

COROLLARY 2. The canonical projection of the fundamental domain D onto X is proper.

If K is a compact subset of E, one has to show that $B = \bigcup_{A \in G} AK \cap D$ is compact. Property (ii) shows that $\bigcup AK$ is locally finite, hence closed; on the other hand, property (i) shows that I(w) is bounded for $w \in B$. The set B, being closed and bounded is therefore compact.

COROLLARY 3. Let R be the equivalence relation induced on D by the equivalence under G (the relation R has been given explicitly in prop. 2). The canonical projection $D \longrightarrow X$ induces an homeomorphism of D/R onto X.

The map $D/R \longrightarrow X$ is bijective (prop.2), continuous, and proper (follows from cor. 2). Since both D/R and X are locally compact, it is a homeomorphism.

After these preliminaries, we define an analytic structure on X in the following way:

Let $p : E \longrightarrow X$ denote the canonical projection. A continuous function f on an open set U of X is said to be holomorphic if $f \circ p$ is holomorphic on $p^{-1}(U)$. One checks easily that the axioms of a complex analytic structure are verified: if $P \in E$ is a point of order e (i.e. the stability group G_P of P is of order e), we can find a local parameter z_P around P such that G_P operates on z_P by multiplication by e-th roots of unity, and $(z_P)^e$ is a local parameter around $p(P) \in X$ [one can take for instance $z_P = (w - P)/(w - \bar{P})$].

Let now $\hat{X} = X \cup \{\infty\}$ be the one point compactification of X; we now want to extend to $\hat{X}$ the analytic structure of X. Let E_1 be the half plane $I(w) > 1$; it follows from formula (1) that the equivalence relation induced on E_1 by G is given by the translations T^n. One has therefore $p(E_1) \approx E_1/T$; but the mapping $w \longrightarrow q = e^{2\pi i w}$ is an analytic isomorphism of E_1/T onto some open disk (minus the origin); putting $q(\infty) = 0$, we then extend q to $p(E_1) \cup \{\infty\}$, and take this function as a local parameter around ∞ on $\hat{X}$. This gives the desired extension. A meromorphic function on $\hat{X}$ may therefore be defined as a meromorphic function on E, invariant under G, admitting a power series expansion $\Sigma_{n \geq k} a_n q^n$ which converges around $q = 0$. Such a function is called a modular function.

PROPOSITION 6. The space $\hat{X}$ is analytically isomorphic to the sphere $S_2 = \hat{C}$.

By cor. 3 to prop. 3, X is homeomorphic to D/R, i.e. to a plane. The Riemann surface $\hat{X}$ is therefore homeomorphic to S_2, and, as is well known, this implies that it is analytically isomorphic to S_2.

If $\lambda : \hat{X} \longrightarrow \hat{C}$ is an analytic isomorphism, every modular function is a rational function of λ. Such a λ is determined up to an analytic automorphism of $\hat{C}$; we normalize it by asking that it maps ∞ and $p(\rho)$ onto ∞ and 0 respectively, and has residue equal to 1 at ∞ (when expanded in a power series of q). This particular choice of λ will be referred to as "the" modular function, and be denoted by j.

Note. Proposition 6 means that X is analytically isomorphic to C. This can also be proved by using the fact that the "half funddamental domain" $D_+ = \{w \mid 0 < Rw < \frac{1}{2}, |w| > 1\}$ is analytically isomorphic to a half plane, and applying Schwarz's symmetry principle.

§4. Modular functions of weight k.

A (meromorphic) modular function of weight k ($k \in \mathbb{Z}$) is a meromorphic function h on E such that:

(2) $$h(A.w) = (cw + d)^{2k} h(w) \qquad (A \in G),$$

and which admits a Laurent expansion in $q = e^{2\pi i w}$ with a finite number of negative exponents:

$$h = \Sigma_{n \geq k} a_n q^n.$$

A function of weight 0 is thus a modular function in the sense of §3. A modular function of weight k which is holomorphic in E and at ∞ is called a modular form.

Equation (2) is equivalent to the fact that the function of w_1, w_2 defined by:

$$h(w_1,w_2) = w_2^{-2k} h(w_1/w_2)$$

is invariant under the group $SL(2,\mathbb{Z})$; note that $f(w_1,w_2)$ is homogeneous of degree -2k.

Since $d(\frac{aw+b}{cw+d}) = dw/(cw+d)^2$, one can also express (2) by saying that the differential form of degree k $\omega = h\, dw^k$ is invariant by G, and hence represents a differential form on $\hat{X}$. The condition that h be meromorphic on E and at ∞ means that ω is everywhere meromorphic on $\hat{X}$. More precisely:

PROPOSITION 7. Let $\omega = h\, dw^k$ be the differential form on $\hat{X}$ associated with the modular function h of weight k. For any point $P \in E$, let $w_P(h)$ denote the order of h at P and $v_P(\omega)$ the order of ω at the corresponding point of $\hat{X}$; let $w_\infty(h)$ and $v_\infty(\omega)$ be defined similarly. Then:

(3) $$w_\infty(h) = v_\infty(\omega) + k$$

(4) $$w_P(h) = e.v_P(\omega) + k(e - 1) \qquad (e = \text{order of } P).$$

Equation (3) follows immediately from the relation $2\pi i dw = \frac{dq}{q}$.
For $P \in E$, choose a local parameter z_P such that the isotropy group G_P operates on z_P by multiplication by e-th roots of unity, and let $t = (z_P)^e$.
We can then write:

$$\omega = u\, t^N dt^k \quad (u \text{ holomorphic} \neq 0,\ N = v_P(\omega)),$$
$$\omega = u\, e^k (z_P)^{Ne + k(e-1)} (dz_P)^k,$$

which yields (4).

On an algebraic curve of genus g, the sum of the orders of a differential form of degree k is equal to $k(2g - 2)$. Here $g = 0$. Hence:

COROLLARY 1. <u>If</u> h <u>is a modular function of weight</u> k, <u>one has</u>:

$$(5) \qquad w_\infty(h) + \tfrac{1}{2} w_i(h) + \tfrac{1}{3} w(h) + \Sigma^* w_P(h) = \frac{k}{6},$$

<u>where</u> Σ^* <u>means that</u> P <u>runs through a set of representatives of the equivalence classes of ordinary points</u> (i.e. points of order 1).

[Formula (5) can also be proved directly by integrating dh/h over the boundary of D.]

Let M_k denote the vector space of modular forms of weight k. If h is a non-zero element of M_k, one has $w_P(h) \geq 0$ for all $P \in E$ and also $w_\infty(h) \geq 0$. Formula (5) then shows that $k \geq 0$. For $k = 0$, the only elements of M_k are the constants; for $k = 1$, formula (5) shows that $M_k = 0$. For $k = 2$, one sees that every non-zero element $h \in M_k$ has a zero only on the orbit of ρ, and that these zeros are simple; if $h' \in M_2$, a suitable linear combination $h' - ah$ has a zero outside this orbit, hence is identically zero; this shows that $\dim. M_2 \leq 1$. An analogous result holds for M_3 with i replacing ρ. We have proved:

COROLLARY 2. a) $M_k = 0$ <u>for</u> $k < 0$ <u>and for</u> $k = 1$.

b) $\dim. M_k \leq 1$ <u>for</u> $k = 2,3$.

c) <u>If</u> h <u>is a non-zero element of</u> M_2 (<u>resp.</u> M_3), <u>it has zeros only on the orbit of</u> ρ (<u>resp. of</u> i), <u>and these zeros are simple</u>.

§5. Eisenstein series.

Let $\Gamma = \mathbb{Z}w_1 + \mathbb{Z}w_2$ be a lattice in $\mathbb{C}$, and let k be an integer, with $k \geq 2$. The Eisenstein series $G_k(\Gamma)$ of degree k associated to Γ is defined by the formula:

$$(6) \quad G_k(\Gamma) = \sum_{u \in \Gamma - \{0\}} \frac{1}{u^{2k}} = \sum_{\substack{m,n \in \mathbb{Z} \\ (m,n) \neq (0,0)}} (mw_1 + nw_2)^{-2k}.$$

The series G_2 and G_3 are identical (up to a constant factor) with the g_2 and g_3 of Weierstrass's theory:

$$(7) \qquad g_2 = 60\, G_2, \quad g_3 = 140\, G_3.$$

To the homogeneous function $G_k(w_1,w_2) = G_k(\Gamma)$ is associated a function $G_k(w)$ of $w = w_1/w_2$ by the formula:

$$(8) \quad G_k(w) = (w_2)^{2k}\, G_k(w_1,w_2) = \sum_{m,n \in \mathbb{Z},\ (m,n) \neq (0,0)} (m+nz)^{-2k}.$$

PROPOSITION 8. <u>For $k \geq 2$, the series (8) is absolutely convergent in E and normally convergent in D. Its sum $G_k(w)$ is a modular form of weight k whose value at infinity is equal to $2\,\zeta(2k)$.</u>

In D, we have $|R(w)| \leq \frac{1}{2}$ and $|w| \geq 1$, whence:

$$|m + nz|^2 \geq m^2 - mn + n^2,$$

which is a positive definite quadratic form. This implies the normal convergence of (8) in D, hence its absolute convergence in E. The fact that $G_k(w_1,w_2)$ is invariant by $SL(2,\mathbb{Z})$ implies that $G_k(w)$ verifies (2). When w tends to infinity in D, each term $(m + nz)^{-2k}$, $n \neq 0$, tends to zero, and $G_k(w)$ tends to $\sum_{m \neq 0} m^{-2k} = 2\,\zeta(2k)$; hence $G_k(w)$ is a modular form.

Let now $\Delta = g_2^3 - 27\, g_3^2$. It is clear that Δ is a <u>modular form of weight</u> 6; since $g_2(\rho) = 0$ and $g_3(\rho) \neq 0$ (cor. 2 to prop. 7), one has $\Delta(\rho) \neq 0$ and Δ is not identically zero. On the other hand:

$$g_2(\infty) = 60\, G_2(\infty) = 120\, \zeta(4) = 120\, \pi^4/90 = \frac{4}{3}\pi^4$$
$$g_3(\infty) = 140\, G_3(\infty) = 280\, \zeta(6) = 560\, \pi^6/945 = \frac{8}{27}\pi^6,$$

hence $\Delta(\omega) = 0$. Formula (5) then shows that $w_\infty(\Delta) = 1$ and that Δ is everywhere $\neq 0$ on E.

PROPOSITION 9. Let $\varepsilon : M_k \longrightarrow \underline{C}$ be the homomorphism $h \longrightarrow h(\infty)$, and $\Delta : M_{k-6} \longrightarrow M_k$ be the multiplication by Δ. For $k \geq 2$ the sequence:

$$0 \to M_{k-6} \xrightarrow{\Delta} M_k \xrightarrow{\varepsilon} \underline{C} \to 0,$$

is exact.

Since $\varepsilon(G_k) = 2\zeta(2k) \neq 0$, ε is surjective. If now $\varepsilon(h) = 0$, then function h/Δ is holomorphic in E (since $\Delta \neq 0$ on E), holomorphic at ∞ (since $w_\infty(\Delta) = 1$), and of weight $k-6$, hence belongs to M_{k-6}.

COROLLARY 1. For $k \geq 2$, the dimension of the space M_k is given by:

$$\dim. M_k = \begin{cases} [k/6] \text{ if } k \equiv 1 \text{ mod. } 6 \\ [k/6] + 1 \text{ if } k \not\equiv 1 \text{ mod. } 6. \end{cases}$$

Proof by induction on k, using cor. 2 to prop. 7.

COROLLARY 2. Every modular form is an isobaric polynomial in g_2 and g_3.

Let $k \geq 2$. There exist positive integers a,b such that $2a + 3b = k$; the form $g_2^b g_3^a$ belongs to M_k, and $\varepsilon(g_2^b g_3^a) \neq 0$. Prop. 9 then shows that every element $h \in M_k$ may be written

$$h = \lambda\, g_2^b g_3^a + \Delta h', \qquad \text{with } h' \in M_{k-6},$$

and our contention follows by induction on k.

Remark. Cor. 1 can also be obtained by the following argument: Let P (resp. Q,R) denote the point at infinity (resp. the image of i, ρ) in $\hat{X}$. Proposition 7 shows that M_k is canonically isomorphic with the vector space Ω_k of differential forms ω of degree k on $\hat{X}$ whose divisors (ω) verify:

(9) $$(\omega) \geq -kP - [k/2]Q - [2k/3]R.$$

If K denotes the canonical class of $\hat{X}$, Ω_k is in turn isomorphic with $L(D_k)$, with $D_k = kK + kP + [k/2]Q + [2k/3]R$. If $k \geq 2$, one has $\deg(D_k) = -k + [k/2] + [k/3] \geq -1$, and the Riemann-Roch formula (applied here to a curve of genus zero!) gives:

$$\dim.M_k = \dim.L(D_k) = \deg(D_k) + 1 = 1 - k + [k/2] + [2k/3],$$

an expression which is easily seen to be equivalent to cor.1.

This interpretation of M_k (which applies as well to other discontinuous groups than the modular group) can also be used to derive explicit formulas for the G_k in terms of j. One gets for instance:

(10) $$g_2 dw^2 = -\frac{\pi^2}{3} \frac{dj^2}{j(j-2^6 3^3)}, \qquad g_3 dw^3 = \frac{\pi^3}{27i} \frac{dj^3}{j^2(j-2^6 3^3)}$$

§6. The q-expansions of G_k, Δ and j.

If n is an integer ≥ 1 and s any number, we put:

$$\sigma_s(n) = \Sigma_{d|n}\, d^s.$$

PROPOSITION 10. The q-expansion of G_k, $k \geq 2$, is:

(11) $$G_k = 2\,\zeta(2k) + 2.(2\pi)^{2k}\,(-1)^k/(2k-1)!\,\Sigma_{n=1}^{\infty}\,\sigma_{2k-1}(n)q^n.$$

We start from the equality:

$$\sum_{m\in\mathbb{Z}} (m+w)^{-2} = \pi^2/\sin^2\pi w = (2\pi i)^2\,\Sigma_{n=1}^{\infty}\, n\, q^n \qquad (q = e^{2\pi i w}).$$

Taking successive derivatives with respect to w, we get:

(12) $$(s-1)!\,\Sigma_{m\in\mathbb{Z}}\,(m+w)^{-s} = (-2\pi i)^s\,\Sigma_{n=1}^{\infty}\, n^{s-1}\, q^n.$$

On the other hand, we have:

$$G_k = 2\,\zeta(2k) + 2\Sigma_{n=1}^{\infty}\,\Sigma_{m\in\mathbb{Z}}\,(m+nw)^{-2k}.$$

Combined with (12), this gives

$$G_k = 2\,\zeta(2k) + 2.(-2\pi i)^{2k}/(2k-1)!\,\Sigma_{\substack{d\geq 1\\ n\geq 1}}\, d^{2k-1}\, q^{dn},$$

hence (11).

PROPOSITION 11. We have $\Delta = (2\pi)^{12} q(1+a_1 q+\dots)$ and $j = q^{-1} + b_o + b_1 q + \dots$ where the a_i's and the b_i's are rational integers. Moreover, j is equal to $2^6 3^3 g_2^3/\Delta$.

Let us put $U = \Sigma_{n=1}^{\infty} \sigma_3(n) q^n$ and $V = \Sigma_{n=1}^{\infty} \sigma_5(n) q^n$. Prop. 10 then shows that:

$$(13) \quad \begin{cases} g_2 = 60\, G_2 = (2\pi)^4 \dfrac{1}{2^2 3} (1 + 240\, U) \\ g_3 = 140\, G_3 = (2\pi)^6 \dfrac{1}{2^3 3^3} (1 - 504\, V) \end{cases}$$

hence:

$$(14) \qquad \Delta = (2\pi)^{12} \frac{1}{2^6 3^3} [(1 + 240\, U)^3 - (1 - 504\, V)^2].$$

The fact that the a_i's are integers is therefore equivalent to the congruence

$$(15) \qquad (1 + 240\, U)^3 \equiv (1 - 504\, V)^2 \qquad \text{mod. } 2^6 3^3.$$

A little calculation shows that (15) is in turn equivalent to the congruence $U \equiv V$ mod. $2^2 3$, or $\sigma_3(n) \equiv \sigma_5(n)$ mod. $2^2 3$; this last congruence is a trivial consequence of $d^3 \equiv d^5$ mod. $2^2 3$. This proves our assertion on Δ.

The function g_2^3/Δ is a modular function which is holomorphic on E and has a simple pole at infinity. Hence $g_2^3/\Delta = aj + b$, with $a,b \in \underline{C}$. Since $g_2(\rho) = j(\rho) = 0$, one has $b = 0$. One the other hand $g_2(\infty) = 4\pi^4/3$, and therefore the coefficient of q^{-1} in the q-expansion of g_2^3/Δ is equal to $2^{-6}3^{-3}$. Since the corresponding coefficient of j is equal to 1 (by definition), this gives $a = 2^{-6}3^{-3}$, and $j = 2^6 3^3 g_2^3/\Delta$.

We then have:

$$j = (1 + 240\, U)^3 q^{-1} (1+a_1 q+\dots)^{-1},$$

and since the a_i's and the coefficients of U are in $\underline{Z}$, so are the coefficients of j, q.e.d.

PROPOSITION 12 ("q-expansion principle"). Let $f = \Sigma_{n \geq -N} x_n q^n$ be a modular function which is holomorphic in E and has a pole of order N at infinity. Then f is a polynomial in j of degree N:

$$f = \Sigma_{n=0}^{n=N} y_n j^n.$$

Moreover, the additive subgroup A of $\underline{C}$ generated by the x_n's is the same as the additive subgroup B generated by the y_n's.

Proof by induction on N. If $N = 0$, f is a constant x_o, and $A = B = \underline{Z}x_o$. If $N > 0$, put $g = f - x_{-N} j^N$. The function g has a pole of order $\leq N-1$ at infinity, and may then be written:

$$g = \Sigma_{n<N} y_n j^n.$$

One has $f = g + x_{-N} j^N$, which shows that $B(f)$ is the subgroup of $\underline{C}$ generated by $B(g)$ and x_{-N}. On the other hand, the fact that j has integral coefficients shows that $A(f)$ is also generated by $A(g) = B(g)$ and x_{-N}, q.e.d.

§7. Explicit formula for the q-expansion of Δ.

This formula will not be needed to prove the main results on complex multiplication outlined in I. However, since it is of considerable interest in itself (Ramanujan's function!), we give here a proof of it. We will essentially follow Hurwitz (Math. Werke, Bd. I, S. 1-67, 578-595).

PROPOSITION 13. One has $\Delta = (2\pi)^{12} q \prod_{n=1}^{\infty} (1-q^n)^{24}$.

The product $f = q \prod_{n=1}^{\infty} (1-q^n)^{24}$ is a holomorphic function on E, invariant by the translation T, and holomorphic at infinity. It is enough to prove it is of weight 6, since prop. 9 will then show that it is equal to $\lambda \Delta$, and the factor λ will be determined by looking at the coefficient of q. By prop.3, we then have to show that

$$(16) \qquad f(-\tfrac{1}{w}) = w^{12} f(w) \qquad \text{for } w \in E.$$

Since f does not vanish on E, one can pick up a determination of $f^{1/24}$, for instance:

$$\eta = e^{\pi i w/12} \prod_{n=1}^{\infty} (1 - e^{2\pi i n w}).$$

The η function is holomorphic on E, and (16) is equivalent to

(17) $$\eta\left(-\frac{1}{w}\right) = \sqrt{w/i}\ \eta(w).$$

The rest of this § is devoted to the proof of (17). For another proof, see for instance C.L. Siegel, Mathematika, 1, p.4.

LEMMA 1. <u>The series</u> $G_1(w) = \Sigma_n [\Sigma_m (m+nw)^{-2}]$ <u>converges absolutely and one has</u> $G_1(w)\, dw = -4\pi i\, d\eta/\eta$.

(As usual, one removes the term corresponding to (n,m) = (0,0)). Using again the formula $\pi^2/\sin^2 \pi n w = \Sigma_m (m+nw)^{-2}$, we get the absolute convergence in m (for fixed n), and the formula:

$$\begin{aligned} G_1(w) &= 2\zeta(2) + 2\Sigma_{n=1}^{\infty} \pi^2/\sin^2 \pi n w \\ &= \pi^2/3 - 8\pi^2 \Sigma_{n=1}^{\infty} q^n/(1-q^n)^2, \end{aligned}$$

whence the absolute convergence in n.

This can be written:

$$G_1(w)dw = (\pi/6i)\ (1 - 24\Sigma_{n=1}^{\infty} \Sigma_{m=1}^{\infty} n\, q^{nm})\ dq/q.$$

On the other hand:

$$\begin{aligned} d\eta/\eta &= dq/24q - \Sigma_{n=1}^{\infty} n\, q^{n-1} dq/(1-q^n) \\ &= (dq/24q)\ (1 - 24\Sigma_{n=1}^{\infty} \Sigma_{m=1}^{\infty} n\, q^{nm}), \end{aligned}$$

whence the lemma.

LEMMA 2. <u>We have:</u>

(18) $$\Sigma_m [\Sigma_n (m-1+nw)^{-1} (m+nw)^{-1}] = -2\pi i/w$$

(19) $$\Sigma_n [\Sigma_m (m-1+nw)^{-1} (m+nw)^{-1}] = 0,$$

<u>where the series are absolutely convergent.</u>

(As before, we remove the terms $(n,m) = (0,0), (0,1)$).

Let denote by H (resp. H_1) the left side of (18) (resp. of (19)). It is clear that the first summations are absolutely convergent. The formula :

$$(m-1 + nw)^{-1} (m+nw)^{-1} = (m-1 + nw)^{-1} - (m+nw)^{-1}$$

gives:

$$\Sigma_n (m-1 +nw)^{-1} (m+nw)^{-1} = \lim_{n \to \infty} \Sigma_{-n}^{n} (m-1 + nw)^{-1} - (m+nw)^{-1}$$

$$= (\pi/w) (\cot\pi\frac{m-1}{w} - \cot\pi\frac{m}{w}),$$

with the convention $\cot a/w = 0$ if $a = 0$.

We then have:

$$H = (\pi/w) \Sigma_m (\cot\pi\frac{m-1}{w} - \cot\pi\frac{m}{w}),$$

a series which is easily proved to be absolutely convergent. Its sum is given by:

$$H = (\pi/w) \lim_{m \to \infty} \Sigma_{-m}^{m} (\cot\pi\frac{m-1}{w} - \cot\pi\frac{m}{w})$$

$$= (\pi/w) \lim_{m \to \infty} (\cot\pi\frac{m+1}{w} + \cot\pi\frac{m}{w}) = - 2\pi i/w,$$

since $\cot\pi\frac{m}{w}$ tends to i when m tends to infinity.

On the other hand, we have:

$$\Sigma_m (m-1 + nw)^{-1} (m+nw)^{-1} = \lim_{m \to \infty} \Sigma_{-m}^{m} ((m-1 + nw)^{-1} - (m+nw)^{-1})$$

$$= \lim_{m \to \infty} ((-m-1 +nw)^{-1} - (m+nw)^{-1}) = 0,$$

whence $H_1 = 0$.

LEMMA 3. <u>We have</u> $G_1(-\frac{1}{w}) = w^2 G_1(w) - 2\pi i w$.

Let us put

$$G(w) = \Sigma_m \Sigma_n (m+nw)^{-2}.$$

We have:

$$H_1 - G_1 = \Sigma_n \Sigma_m (m+nw)^{-2} (m-1 + nw)^{-1},$$

which is absolutely convergent (as a <u>double</u> series). We can then interchange the order of summations, and we get:

$$H_1 - G_1 = H - G.$$

Using lemma 2, we have then $G_1(w) - G(w) = 2\pi i/w$. Using the obvious formula:

$$G_1(-\frac{1}{w}) = w^2 G(w),$$

we obtain lemma 3.

Proof of (17).

Lemma 3 may be written:

$$G_1(-\frac{1}{w}) . d(-\frac{1}{w}) = G_1(w)dw - 2\pi i\, dw/w.$$

Using lemma 1, this gives:

$$(d\eta/\eta)(-\frac{1}{w}) = d\eta/\eta + dw/2w,$$

which means that

$$\eta(-\frac{1}{w}) = C\, w^{1/2}\, \eta(w).$$

Putting $w = i$ gives $C\sqrt{i} = 1$, and (17) is proved.

III CLASS INVARIANTS I

(A. Borel, Nov. 6, 1957)

§1. Introduction.

Let $X = \underline{C}/\Gamma$ be an elliptic curve. If X has a non-trivial complex multiplication, the ring A(x) of all complex multiplications of X forms an order of an imaginary quadratic field. Let an imaginary quadratic field K be fixed. Then there is a one-one correspondence between the set of ideal classes $\underline{k}_1, \ldots, \underline{k}_h$ of K and the set of X such that A(x) coincides with the ring of all algebraic integers in K; the correspondence is given by assigning to $\underline{k}_i$ the elliptic curve with the invariant $j(\underline{k}_i)$. (Cf. I §1.)

The main purpose of the four following lectures is to prove that the class invariants $j(\underline{k}_i)$ are all algebraic integers forming a full set of conjugates over K, and also over **Q**, and that $K(j(\underline{k}_i))$ is the maximal unramified abelian extension of K. As mentioned in I, we follow Hasse's first method, which uses analysis and class field theory.

§2. Modular correspondences.

Let E be the upper-half complex plane, G the modular group and S the closed Riemann surface obtained by adjoining ∞ to E/G. S is mapped isomorphically onto the complex sphere by the function j (cf. II §3). So S is an algebraic curve of genus 0 and an algebraic correspondence of S into itself is called a modular correspondence.

A method to generate such a modular correspondence is as follows: Let $\tilde{G}$ be the group of all automorphisms of the Riemann surface E. It is the quotient of the group GL(2, R) of real 2 x 2 matrices by its center. Take a subset H of $\tilde{G}$ containing G such that GH = H, $[H : G] = m < \infty$ and $H^{-1} = H$ (hence also HG = H). For any M in H, let $j_M(z)$ denote the function $j(M(z))$. Then j_M depends only upon the class GM of M in H/G. Let $j_1(z), \ldots, j_m(z)$ be the functions obtained in this way from all m classes of H/G. Then it can easily be proved that each element of H may be represented by a matrix with integral rational coefficients and that each class GM contains a triangular matrix. It follows then (see the proof of Theorem 1a below) that the symmetric functions of $j_1, \ldots, j_m$ are integral modular functions, i.e. polynomials in j. Hence, $F(t, j) = \prod_{s=1}^{m} (t - j_s(z))$ is a polynomial of t over $\underline{C}(j)$, and a mapping which maps a point of S with "coordinate" j(z) to the points with coordinates $j_1(z), \ldots, j_m(z)$ defines a modular correspondence.

Taking any modular function g(z) instead of j(z), we can define $g_1(z), \ldots, g_m(z)$ just as above, and symmetric functions of $g_1, \ldots, g_m$ then give us modular functions. More generally, let h be any modular form of weight k. For M in H, put $h_M(z) = h(M(z))(\frac{dM(z)}{dz})^k$. Then $h_M(z)$ again depends only upon the class GM and we get m such functions $h_1(z), \ldots, h_m(z)$ from m classes of H/G. A symmetric function of weight ν in $h_1, \ldots, h_m$ is then a modular form of weight νk.

§3. The correspondences F_n.

We now consider a special kind of correspondences. Let n be any positive integer. We denote by H_n^* the set of all automorphisms of E given by

$$M : z \longrightarrow \frac{az + b}{cz + d}, \qquad ad - bc = n, \quad a, b, c, d \in \underline{Z},$$

and denote by H_n the subset of H_n^* consisting of those M satisfying

$(a, b, c, d) = 1$. Clearly, $GH_n^* = H_n^*$, $H_n^{*-1} = H_n^*$ and $GH_n = H_n$, $H_n^{-1} = H_n$. Furthermore, a set of representatives for classes GM of H_n^*/G is obtained by the automorphisms with matrices

$$(1) \qquad \begin{pmatrix} a & b \\ 0 & d \end{pmatrix}, \quad a > 0,\ d > 0,\ d > b \geq 0,\ ad = n .$$

Hence, $[H_n^* : G] = \sigma_1(n) = \sum_{d|n} d$. Imposing a further condition $(a, b, d) = 1$ on matrices in (1), we get a set of representatives for H_n/G and it is not difficult to see that $[H_n : G] = \psi(n) = n \prod_{p|n} (1 + \frac{1}{p})$.

Another way of defining H_n is as follows: Let Γ be a two-dimensional lattice on the complex plane. A sublattice Γ' of Γ with index n is called primitive if Γ/Γ' is a cyclic group of order n. Let ω_1, ω_2 be a basis of Γ and ω_1', ω_2' a basis of such a primitive sublattice Γ'. Put $\omega_1' = a\omega_1 + b\omega_2$, $\omega_2' = c\omega_1 + d\omega_2$. Then $M : z \longrightarrow \frac{az + b}{cz + d}$ is in H_n. Conversely, every M in H_n can be obtained in this way from some Γ, Γ' and their suitable bases ω_1, ω_2 and ω_1', ω_2'.

Thus if X is a curve with lattice of periods, Γ the correspondence F_n defined by H_n associates to X the curves having a lattice of periods primitive of of index n in Γ. Analogously, H_n^* associates to X all curves which have a period lattice of index n in Γ.

Given such Γ' primitive of index n in Γ, we can always find, by a theorem on abelian groups, a basis ω_1, ω_2 of Γ such that $n\omega_1$, ω_2 form a basis of Γ'. Hence, if M_o denotes the automorphism $z \longrightarrow nz$, then $H_n = GM_oG$. Therefore, right multiplication of G on H permutes the classes GM of H/G <u>transitively</u>. In general, this is not true for H_n^*.

§4. The modular equations.

Now, let $M_1, \ldots, M_N (N = \psi(n))$ be the representatives for H_n/G given by the matrices (1) with $(a, b, d) = 1$. Put

$$j_s(z) = j(M_s(z)) , \qquad s = 1, \ldots, N .$$

These are the functions associated with N classes of H_n/G and, as $H_n G = H_n$, $j_s(z) \longrightarrow j_s(T(z))$ is a permutation of $j_1, \ldots, j_N$ for any T in G. Hence, a symmetric function of $j_1, \ldots, j_N$ is invariant under $z \longrightarrow T(z)$, $T \in G$.

THEOREM 1. (a) $F_n(t, j) = \prod_{s=1}^{N} (t - j_s(z))$ is a polynomial of t and $j = j(z)$ with integral rational coefficients,

(b) if n is not a square, the highest coefficient of j in $F_n(j, j)$ is ± 1,

(c) $F_n(t, j)$ is an irreducible polynomial of t over the field $\underline{C}(j)$,

(d) $F_n(t, j) = F_n(j, t)$, $n > 1$.

Proof. (a) An elementary symmetric function $\sigma_\nu (j_1(z), \ldots, j_N(z))$ of $j_1, \ldots, j_N$ is, as noticed above, invariant under G and is obviously a holomorphic function of z in E. To see the behavior of σ_ν at ∞, put $q = e^{2\pi i z}$. Then, by II §6, $j(z) = q^{-1}(1 + A(q))$ where $A(q)$ is a power series of q with integral rational coefficients and $A(0) = 0$. Let

$$M_s \longleftrightarrow \begin{pmatrix} a_s & b_s \\ 0 & d_s \end{pmatrix}, \quad e^{2\pi i M_s(z)} = \zeta_{d_s}^{b_s} q^{\frac{a_s}{d_s}}, \quad \zeta_{d_s} = e^{\frac{2\pi i}{d_s}} .$$

Then,

$$(2) \qquad j_s(z) = j(M_s(z)) = \zeta_{d_s}^{-b_s} q^{-\frac{a_s}{d_s}} (1 + A(\zeta_{d_s}^{b_d} q^{\frac{a_s}{d_s}})) .$$

Therefore, σ_ν can only have the singularity of a pole in q at q = 0, and, hence, is a polynomial in j over $\underline{\underline{C}}$. Furthermore, the coefficients of the q-expansion of σ_ν are all algebraic integers in the field $Q(\zeta_n)$, $\zeta_n = e^{\frac{2\pi i}{n}}$. Let τ be any Galois automorphism of $Q(\zeta_n)/Q$. Then $\tau(\zeta_n) = \zeta_n^f$ for some f with (n, f) = 1, and we can write

$$\tau(\zeta^{\frac{b_s}{d_s}}) = \zeta^{\frac{fb_s}{d_s}} = \zeta^{\frac{b_t}{d_t}} ,$$

where

$$\begin{pmatrix} a_t, & b_t \\ 0 , & d_t \end{pmatrix} , \quad a_t = a_s, \; d_t = d_s, \; b_t \equiv fb_s \bmod. d_s ,$$

is another matrix in (1) uniquely determined by M_s. Hence, $\tau(j_s) = j_t$ and τ permutes the functions $j_1, \ldots, j_N$ among themselves. Thus, $\tau(\sigma_\nu) = \sigma_\nu$ for any τ in the Galois group of $Q(\zeta_n)/Q$, and the coefficients of the q-expansion of σ_ν must be rational integers. By the q-expansion principle (II §6), σ_ν is then a polynomial in j with integral rational coefficients.

(b) Since n is not a square, $\frac{a_s}{d_s}$ in (2) cannot be 1. Hence, the leading coefficient of the q-expansion of $j - j_s$ is a root of unity, and so is the leading coefficient of the q-expansion of the product $F_n(j, j) = \prod_{s=1}^{N} (j - j_s)$. However, this coefficient is equal to the highest coefficient of j in $F_n(j, j)$ and, as it is rational, it must be ± 1.

(c) This follows immediately from the fact that $j_1, \ldots, j_N$ are permuted transitively among themselves under $j_s(z) \longrightarrow j_s(T(z))$, $T \in G$. (Cf. the last lines of §3.)

(d) Since $z \longrightarrow nz$ and $z \longrightarrow \frac{z}{n}$ are both contained in H_n, $F_n(j(nz), j(z)) = 0$ and $F_n(j(\frac{z}{n}), j(z)) = 0$, identically in $z \in E$. Replacing z by nz in the second equality, we get $F_n(j(z), j(nz)) = 0$. Hence, as polynomials in $\underline{C}(j)[t]$, $F_n(t, j)$ and $F_n(j, t)$ have a common zero $t = j(nz)$. Since $F_n(t, j)$ is irreducible by c) and since the highest coefficient of t in $F_n(t, j)$ is 1, we obtain

$$F_n(j, t) = P(t, j)F_n(t, j) ,$$

with a polynomial $P(t, j)$ in $\underline{Z}[t, j]$. It then follows that

$$F_n(t, j) = P(j, t)F_n(j, t) = P(j, t)P(t, j)F_n(t, j) ,$$

and hence, that

$$P(t, j) = P(j, t) = \pm 1.$$

If $P = -1$, we would have $F_n(j, t) = - F_n(t, j)$ and $F_n(j, j) = 0$ so that $F_n(t, j)$ is divisible by $t - j$, contradicting c). Hence, $P = 1$ and $F_n(t, j) = F_n(j, t)$.

$F_n(t, j) = 0$ is the modular equation for the degree n. (For a given $j(z)$, which is the invariant of an elliptic curve X, the roots of $F_n(t, j) = 0$ are the modular invariants of the curves X_i $(1 \leq i \leq N)$ which the correspondence F_n associates to X. Theorem 1d means that then X is among the images of X_i; this can also be seen geometrically by noticing that if Γ' is primitive of index n in Γ, then $n\Gamma$ is primitive of index n in Γ', and that the curves with periods Γ and $n\Gamma$ are isomorphic.

§5. Class invariants.

We next show that an elliptic curve X has non-trivial complex multiplications if and only if the invariant $j(X)$ satisfies $F_n(j(X), j(X)) = 0$ for some $n > 1$. Let $X = \underline{C}/\Gamma$ and let ω_1, ω_2 be a basis of Γ so that $j(X) = j(\omega)$, $\omega = \frac{\omega_1}{\omega_2}$. Suppose, first, that $F_n(j(\omega), j(\omega)) = 0$, $n > 1$. Then $j_s(\omega) = j(\omega)$ for some s and $M(\omega) = \omega$ for some M in H_n. Hence, there exists a w in $\underline{C}$ such that

$$w\omega_1 = a\omega_1 + b\omega_2 , \qquad M \longleftrightarrow \begin{pmatrix} a & b \\ c & d \end{pmatrix} .$$
$$w\omega_2 = c\omega_1 + d\omega_2 ,$$

Then, w cannot be a rational integer and X has a non-trivial complex multiplication. Conversely, suppose X has non-trivial complex multiplications. Then the ring of complex multiplications $A(X)$ is an order of an imaginary quadratic field K and we can find a w in $A(X)$ such that $N_{K/Q}(w) = n > 1$ and that w is not in any $mA(X)$, $m > 1$. Reversing the above argument, we see that $j(X)$ satisfies $F_n(j(X), j(X)) = 0$. (Cf. below.)

We now fix an imaginary quadratic field K and consider the class invariants $j(\underline{k}_i)$, $i = 1, \ldots, h$, of K, i.e. the invariants $j(X)$ of elliptic curves X of which $A(X)$ coincide with the ring of all algebraic integers in K. Let w be an algebraic integer in K such that $n = N_{K/Q}(w)$ is a square-free integer > 1. Such a w always exists; if $K = Q(\sqrt{-1})$, take $w = 1 + \sqrt{-1}$ and if $K = Q(\sqrt{-m})$, $m > 1$ and square-free, take $w = \sqrt{-m}$. Let ω_1, ω_2 be a basis of an ideal in an ideal class $\underline{k}$ of K and put

$$w\omega_1 = a\omega_1 + b\omega_2 ,$$
$$w\omega_2 = c\omega_1 + d\omega_2 , \qquad a, b, c, d \in \underline{Z} .$$

Then $ad - bc = N_{K/Q}(w) = n$ and, as n is square-free, $(a, b, c, d) = 1$. Hence, $M : z \longrightarrow \frac{az + b}{cz + d}$ is in H_n. Since $j(M(\omega)) = j(\omega)$, the invariant $j(\underline{k}) = j(\omega)$

then satisfies $F_n(j(\underline{k}), j(\underline{k})) = 0$, and from the previous theorem, we obtain immediately the following:

THEOREM 2. *The class invariants* $j(\underline{k}_i)$, $i = 1, \ldots, h$, *of an imaginary quadratic field are algebraic integers.*

IV CLASS INVARIANTS II

(A. Borel, Nov. 13, 1957)

§1. Introduction.

The main purpose of this lecture is to obtain Theorem 3. For this we shall first establish some properties of certain functions formed by means of the discriminant Δ (see §2, 3). We follow the usual conventions of algebraic number theory in which often no distinction is made between an ideal $\underline{a}$ in an algebraic number field K and the ideal $\underline{o}(L).\underline{a}$ it generates in a finite extension L of K. In particular let $\underline{a}_1$, $\underline{a}_2$ be ideals in K_1, K_2 and L be a finite extension of K_1 and K_2. Then we say that $\underline{a}_1$ divides $\underline{a}_2$ (resp. that $\underline{a}_1$ and $\underline{a}_2$ are prime to each other) if $\underline{o}(L).\underline{a}_1$ divides $\underline{o}(L).\underline{a}_2$ (resp. $\underline{o}(L).\underline{a}_1$ and $\underline{o}(L).\underline{a}_2$ are prime to each other). This is then true in any algebraic extension of K_1 and K_2.

§2. The functions φ_M.

In dealing with Δ, it will be convenient to use the homogeneous formulation. We denote in the same way the automorphism of the upper half plane E given by

$$z \longrightarrow (az + b)(cz + d)^{-1}$$

and the homogeneous linear transformation $(w_1, w_2) \longrightarrow (aw_1+bw_2, cw_1+dw_2)$. We recall that if h(z) is a modular form of weight k, then

$$h(w_1, w_2) = w_2^{-2k}\, h(w_1/w_2)$$

is a homogeneous function of degree -2k in w_1, w_2 and is invariant under the "homogeneous" modular group G. In particular, the modular form Δ

gives rise to a homogeneous function of degree -12, invariant under G:

$$\Delta(w_1, w_2) = w_2^{-12}\, \Delta(\frac{w_1}{w_2})$$

and we have

(1) $$\Delta(w_1, w_2) = (\frac{2\pi}{w_2})^{12} q(1 + B(q)) , \qquad q = e^{2\pi i \frac{w_1}{w_2}} ,$$

where B(q) is a power series of q with integral rational coefficients and $B(o) = 0$ (see II, §6).

For any M in H_n, put

$$\varphi_M(w_1, w_2) = n^{12} \frac{\Delta(M(w_1, w_2))}{\Delta(w_1, w_2)} .$$

Then φ_M depends only upon the class GM of H_n/G and, from the $N = \psi(n)$ classes of H_n/G, we obtain N homogeneous functions of degree 0 in w_1, w_2:

$$\varphi_i(w_1, w_2) = \varphi_{M_i}(w_1, w_2) , \qquad 1 \leq i \leq N .$$

These functions are regular for $\mathrm{Im}(\frac{w_1}{w_2}) > 0$ and are permuted among themselves under the transformations in G.

Now, it follows immediately from (1) that

(2) $$\varphi_i(w_1, w_2) = a_i^{12} \zeta_{d_i}^{b_i} q^{\frac{a_i}{d_i} - 1}(1 + B(\zeta_{d_i}^{b_i} q)(1 + B(q))^{-1}, \left(\zeta_d = e^{\frac{2\pi i}{d}}\right),$$

and we see, by a similar argument as in III, §4, that

$$\Phi_n(t, j) = \prod_{i=1}^{N} (t - \varphi_i(w_1, w_2))$$

is a polynomial in t and j with integral rational coefficients.

LEMMA 1. __If__ $n = p$ __is a prime,__ $\prod_{i=1}^{N} \varphi_i(w_1, w_2)$ __is a constant and is equal to__ $(-1)^{p-1}p^{12}$.

Proof. In this case, $N = \psi(p) = p+1$ and the matrices M_i are given by

$$M_i = \begin{pmatrix} 1 & i \\ o & p \end{pmatrix}, (1 \leq i \leq p); \quad M_{p+1} = \begin{pmatrix} p & o \\ o & 1 \end{pmatrix}.$$

Hence,

$$\varphi_i(w_1, w_2) = \zeta_p^{i} q^{\frac{1}{p} - 1} (1 + B(\zeta_p^{i} q^{\frac{1}{p}}))(1 + B(q))^{-1}, \quad (1 \leq i \leq p),$$

(3) $$\varphi_{p+1}(w_1, w_2) = p^{12} q^{p-1}(1 + B(q^p))(1 + B(q))^{-1}, \quad (i = p + 1),$$

and the q-expansion of $\prod_{i=1}^{N} \varphi_i(w_1, w_2)$ starts with the constant term $(-1)^{p-1}p^{12}$. However, as this product is also a polynomial in j, it must be equal to the constant $(-1)^{p-1}p^{12}$.

§3. Properties of the singular values of the φ_M.

We now fix an imaginary quadratic field K.

LEMMA 2. Let $\underline{a}$ be an ideal of K and (α_1, α_2) a basis of $\underline{a}$ such that $\mathrm{Im}(\alpha_1/\alpha_2) > 0$. Then, for any M in H_n, $\varphi_M(\alpha_1, \alpha_2)$ is an algebraic integer and, if $n = p$ is a prime, the principal ideal $(\varphi_M(\alpha_1, \alpha_2))$ is a divisor of (p^{12}).

Proof. By the definition of Φ_n,

$$\Phi_n(t, j(\underline{a})) = \prod_{i=1}^{N} (t - \varphi_i(\alpha_1, \alpha_2)).$$

But, as $j(\underline{a})$ is an algebraic integer (III, Theorem 2), the left hand side is a unitary polynomial in t with integral algebraic coefficients. Since $\varphi_M(\alpha_1, \alpha_2)$, ($= \varphi_i(\alpha_1, \alpha_2)$ for some i), is a root of $\Phi_n(t, j(\underline{a})) = 0$, it is an algebraic integer. The second part is then an immediate consequence of Lemma 1.

Let $\underline{a}$ be an integral ideal of K. Then, for any (fractional or integral) ideal $\underline{b}$ of K, $\underline{a}.\underline{b}$ is an additive subgroup of $\underline{b}$, of index equal to $N\underline{a}$, as follows from the fact that $N(\underline{a}.\underline{b}) = N(\underline{a}).N(\underline{b})$. Thus, if (β_1, β_2) is a base of $\underline{b}$, and $(a\beta_1+b\beta_2, c\beta_1+d\beta_2)$ a base of $\underline{a}.\underline{b}$, where $a,b,c,d\in\mathbb{Z}$, we must have $ad-bc = N\underline{a}$; moreover $(a,b,c,d) = 1$ if $\underline{a}$ is not divisible by a rational integer > 1, for instance if $\underline{a}$ is prime.

A prime ideal $\underline{p}$ of K is said to be of first degree if every integer of K is congruent mod $\underline{p}$ to a rational number or equivalently if $p = N\underline{p}$ is a prime number; $\underline{p}$ is unramified for $K/\mathbb{Q}$ if its square does not divide a prime rational number or equivalently if $N\underline{p}$ does not divide the discriminant of $K/\mathbb{Q}$. Under those two conditions we have $\underline{p} \neq \bar{\underline{p}}$ ($^-$ denoting complex conjugation) and $(p) = \underline{p}.\bar{\underline{p}}$.

THEOREM 1. *Let* (α_1, α_2) *be a basis of an ideal* $\underline{a}$ *of* K, $\underline{p}$ *be a prime ideal of the first degree in* K, *unramified for* $K/\mathbb{Q}$, *and* $p = N\underline{p}$. *Let* $P\in H_p$ (*resp.* $\bar{P}\in H_p$) *be such that* $P(\alpha_1, \alpha_2)$ (*resp.* $\bar{P}(\alpha_1, \alpha_2)$) *is a basis of* $\underline{p}.\underline{a}$ (resp. $\bar{\underline{p}}.\underline{a}$). *Then*

$$(\varphi_P(\alpha_1, \alpha_2)) = \bar{\underline{p}}^{12}, \qquad (\varphi_{\bar{P}}(\alpha_1, \alpha_2)) = \underline{p}^{12}$$

and if $M\in H_p$, $M\notin GP, G\bar{P}$, *then* $\varphi_M(\alpha_1, \alpha_2)$ *is a unit.*

Proof. Let f be a positive integer such that $\underline{p}^f$ is a principal ideal in K. We have then $\underline{p}^f = (\alpha)$, and $\alpha\bar{\alpha} = N_{K/\mathbb{Q}}(\alpha) = p^f$, where α is an integer of K. By the discussion preceding Theorem 1, the elements $P, \bar{P}$ are in H_p, and we can find $P_1 = P, P_2, \ldots, P_f$ in H_p such that $P_iP_{i-1} \ldots P_1(\alpha_1, \alpha_2)$ is a basis of $\underline{p}^i.\underline{a}$ $(1 \leq i \leq f)$ and, in particular, $P_f \ldots P_1(\alpha_1, \alpha_2) = (\alpha\alpha_1, \alpha\alpha_2)$. Put

$$\lambda_i = \varphi_{P_i}(P_{i-1}\ldots P_1(\alpha_1, \alpha_2)) = p^{12}\,\frac{\Delta(P_i\ldots P_1(\alpha_1,\alpha_2))}{\Delta(P_{i-1}\ldots P_1(\alpha_1,\alpha_2))}, (1 \leq i \leq f).$$

Then,

$$\prod_{i=1}^{f} \lambda_i = p^{12f} \frac{\Delta(P_f \dots P_1(\alpha_1, \alpha_2))}{\Delta(\alpha_1, \alpha_2)} = p^{12f} \frac{\Delta(\alpha\alpha_1, \alpha\alpha_2)}{\Delta(\alpha_1, \alpha_2)} = p^{12f}\alpha^{-12} = \bar{a}^{-12} .$$

As every λ_i is an algebraic integer by Lemma 2, the ideal (λ_i) divides $(\bar{\alpha}^{12}) = \bar{\underline{p}}^{12f}$. But, by the same lemma, (λ_i) also divides $(p^{12}) = \underline{p}^{12}.\bar{\underline{p}}^{12}$. Hence, (λ_i) is a divisor of $\bar{\underline{p}}^{12} = (\bar{\alpha}^{12}, p^{12})$. Since $\prod_{i=1}^{f}(\lambda_i) = (\bar{\alpha}^{12}) = \bar{\underline{p}}^{12f}$, it follows that $(\lambda_i) = \bar{\underline{p}}^{12}$ for every i, and, in particular, $(\varphi_P(\alpha_1, \alpha_2)) = (\lambda_1) = \bar{\underline{p}}^{12}$. Similarly, $(\varphi_{\bar{P}}(\alpha_1, \alpha_2)) = \underline{p}^{12}$. As $\underline{p} \neq \bar{\underline{p}}$, this shows that $GP \neq G\bar{P}$, and it follows from

$$\prod_{i=1}^{N}(\varphi_i(\alpha_1, \alpha_2)) = (p^{12}) = \underline{p}^{12}\bar{\underline{p}}^{12} ,$$

that $\varphi_M(\alpha_1, \alpha_2)$ is a unit if $GM \neq GP, G\bar{P}$.

§4. A formal congruence.

Let ξ and η be power series (in particular polynomials) of certain variables with integral algebraic coefficients and $\underline{a}$ an ideal of integral numbers. As usual we write

$$\xi \equiv \eta \bmod \underline{a}$$

if every coefficient of $\xi - \eta$ is in $\underline{a}$.

We consider 2(p+1) q-series $\xi_i(q)$, $\eta_i(q)$, $(1 \leq i \leq p+1)$, with the following properties

(i) Their coefficients are integral numbers of $\mathbb{Q}(\zeta_p)$; ξ_p, ξ_{p+1} η_p, η_{p+1} have rational integral coefficients. The ξ_i's (resp. η_i's) $(1 \leq i \leq p-1)$ are permuted by the automorphisms of the Galois group of $\mathbb{Q}(\zeta_p)/\mathbb{Q}$, and $\xi_i \equiv \xi_j$, $\eta_i \equiv \eta_j \bmod (1-\zeta_p)$, $(1 \leq i, j \leq p)$.

(ii) They represent holomorphic functions in the upper half plane which are permuted among themselves by G.

By (4) the φ_i's satisfy these assumptions. The same is true for the j_i's because we have

$$(4) \qquad \begin{aligned} j_i(q) &= \zeta_p^{-i} q^{-1/p}(1 + A(\zeta_p^{i} q^{1/p})) , \\ j_{p+1}(q) &= q^{-p}(1 + A(q^p)) \end{aligned}$$

where $A(q)$ has rational integral coefficients. In fact, these will be the only functions satisfying (i)(ii) to be considered in the applications.

With indeterminates t, u we then put

$$(5) \qquad G_p(t, u, \xi_i, \eta_j) = \Sigma_{j=1}^{j=p+1}[(t-\xi_j)\prod_{k=1,k\neq j}^{k=p+1}(u-\eta_k)] .$$

LEMMA 4. (a) $G_p(t, u, \xi_i, \eta_j) = \tilde{G}_p(t,u,j)$ <u>is a polynomial in t,u,j with rational integral coefficients.</u>

(b) $G_p(t, u, \xi_i(q), \eta_j(q)) \equiv (t-\xi_{p+1}(q))(u^p - \eta_p(q)^p) \bmod p$.

(a) is proved exactly as Theorem 1a of III. Since in $Q(\zeta_p)$, the ideal (p) is equal to $(1-\zeta_p)^{p-1}$ and since both sides of (b) have rational integral coefficients it is enough to show that both sides of (b) are congruent mod $(1-\zeta_p)$. By (i) above, we have $\xi_i \equiv \xi_j$ and $\eta_i \equiv \eta_j \bmod (1-\zeta_p)$ for $(1 \leq j, i \leq p)$. Therefore

$$G_p \equiv (t-\xi_{p+1})(u-\eta_p)^p + p(u-\eta_{p+1})(t-\xi_p)(u-\eta_p)^{p-1} \bmod (1-\zeta_p) ,$$

and our contention follows from the fact that mod p, we have $(u-\eta_p)^p \equiv u^p - \eta_p^p$.

§5. <u>Applications</u>.

The modular equation for the degree p is $F_p(t,j) = 0$ where

$F_p(t,j) = \prod(t-j_i)$ is a polynomial in t and j with rational integral coefficients (III Theorem 1a). The following property of F_p is a classical result due to H. Weber (Acta Mathematica 6, 1885, p. 390).

THEOREM 2. We have $F_p(t,j) \equiv (t-j^p)(t^p-j) \bmod p$, the integer p being prime.

From (4) it follows by Fermat's first theorem that

$$(6) \qquad \begin{aligned} j_{p+1}(q) &\equiv j(q)^p && \bmod p \\ j_p(q)^p &\equiv j(q) && \bmod p \end{aligned}$$

In order to apply Lemma 4, we put $t = u$, $\xi_i = \eta_i = j_i$ $(1 \leq i \leq p+1)$. Then $\tilde{G}_p = F_p(t,j)$, and, in view of (6), the right hand side in Lemma 4b is congruent to $(t-j^p)(t^p-j) \bmod p$. Lemma 4 implies therefore that the difference of the two sides in the formula of Theorem 2, when viewed as a power series in t and q, has all its coefficients in $p\mathbb{Z}$. By the q-expansion principle (II §6), the same is then true when this difference is written as a polynomial in t and j.

THEOREM 3. Let K be an imaginary quadratic field, $\underline{p}$ a prime ideal of K of degree 1, unramified for $K/\mathbb{Q}$, and $\underline{k}_{\underline{p}}$ the ideal class containing $\underline{p}$. Then for any ideal $\underline{k}$ of K, we have

$$(7) \qquad j(\underline{k}_{\underline{p}}^{-1}\cdot\underline{k}) = j(\underline{k})^{N\underline{p}} \qquad \bmod \underline{p}\ .$$

Let $\underline{a}$ be an ideal of $\underline{k}$. Since $\underline{p}\cdot\bar{\underline{p}} = (N\underline{p})$, the ideal class $\underline{k}_{\underline{p}}^{-1}$ contains $\bar{\underline{p}}$ and our contention is equivalent to

$$(8) \qquad j(\underline{p}\cdot\underline{a}) \equiv j(\underline{a})^p \bmod \underline{p} \ (p = N\underline{p})\ .$$

Here p is prime, and $\underline{p} \neq \bar{\underline{p}}$. Let α_1, α_2 be a base of $\underline{a}$ and $P, \bar{P} \in H_p$ be as in Theorem 1. Furthermore, let $c,d \leq p+1$ be the indices such that

$P \in GM_c$, $\bar{P} \in GM_d$. Then $c \neq d$ since $\underline{p} \neq \underline{\bar{p}}$ and

(9)
$$j(\underline{\bar{p}} \cdot \underline{a}) = j_{\bar{p}}(\underline{a}) = j_d(\underline{a}) = j_d(\alpha_1/\alpha_2) .$$

We now apply Lemma 4 to the case where $\xi_i = j_i$, $\eta_i = \varphi_i (1 \leq i \leq p+1)$ and $t = j^p$. This gives

$$G_p(j(q)^p, u, j_i(q), \varphi_k(q)) \equiv (j^p - j_{p+1}) \cdot (u^p - \varphi_p^p) \quad \text{mod } p$$

which, together with (6), shows that

$$G_p(j(q)^p, u, j_i(q), \varphi_k(q)) \equiv 0 \text{ mod } p ,$$

G_p being considered as a power series in u and q. By the q-expansion principle (II §6), we have then

(10)
$$\tilde{G}_p(j^p, u, j) \equiv 0 \text{ mod } p$$

where (cf. Lemma 4), $\tilde{G}_p$ is G_p considered as a polynomial in j and u.

Let now $q_o = \exp.(2\pi i \alpha_1/\alpha_2)$. The numbers $j(q_o)$, $j_i(q_o)$, $\varphi_i(q_o)$ are algebraic integers. Since $G_p(j^q, u, j_i, \varphi_j)$ may be written as a __polynomial__ in t and j with coefficients in $p\underline{Z}$, its value for $q = q_o$ is a polynomial in u, whose coefficients are algebraic integers divisible by p. Thus

(11)
$$G_p(j(\underline{a})^p, u, j_i(\underline{a}), \varphi_k(\alpha_1, \alpha_2)) \equiv 0 \text{ mod } p .$$

Let us now put $u = \varphi_{\bar{p}}(\alpha_1, \alpha_2) = \varphi_d(\alpha_1, \alpha_2)$. Then by (11) and the definition of G_p, we get

(12)
$$(j(\underline{a})^p - j_d(\underline{a})) \prod_{i \neq d} (\varphi_d(\alpha_1, \alpha_2) - \varphi_i(\alpha_1, \alpha_2)) \equiv 0 \text{ mod } p .$$

Since $\underline{p}$ divides (p), we have a fortiori

(13) $$(j(\underline{a})^p - j_d(\underline{a})) \prod_{i \neq d} (\varphi_d(\alpha_1, \alpha_2) - \varphi_i(\alpha_1, \alpha_2)) \equiv 0 \bmod \underline{p} .$$

By Theorem 1,

$$\prod_{i \neq d} (\varphi_d(\alpha_1, \alpha_2) - \varphi_i(\alpha_1, \alpha_2)) \equiv (-1)^p \prod_{i \neq d} \varphi_i(\alpha_1, \alpha_2) \bmod \underline{p}$$

and the right hand side generates the ideal $\bar{\underline{p}}^{12}$, which is prime to $\underline{p}$. Therefore, (13) implies that

$$j(\underline{a})^p - j_d(\underline{a}) \equiv 0 \bmod \underline{p}$$

and this, in view of (9), is the congruence (8).

<u>Remarks</u>. (1) We have $j_{p+1}(z) = j(p.z)$, so that the first congruence in (6) is a congruence between the q-developments of $j(z)$ and $j(pz)$, that is between $j(q)$ and $j(q^p)$. It may be viewed as a formal analogue to Theorem 3. In fact, for a suitable base α_1, α_2 of $\underline{a}$ the numbers $p\alpha_1$, α_2 form a base of $\bar{\underline{p}}.\underline{a}$ so that (8) may be written

(14) $$j(\underline{a})^p \equiv j_{p+1}(\underline{a}) \bmod \underline{p} .$$

However, one cannot of course simply derive (14) out of (6) by putting $q = q_o = \exp(2\pi i\alpha_1/\alpha_2)$, because both sides are infinite series in q (and as a matter of fact, this would lead to a congruence mod p, not only mod $\underline{p}$, which is not true in general.) However, a formal congruence between q-series with coefficients in an additive group H of algebraic integers leads to a correct congruence for $q = q_o$ if both sides represent integral modular functions, because they can then be considered as polynomials in j with coefficients in H by the q-expansion principle, and $j(q_o)$ is an algebraic integer. It is this fact which has allowed us to derive (11) from Lemma 4.

(2) M. Eichler, Math. Zeitshrift 64 (1956), 229-42 has given a proof of the main theorems of complex multiplication which uses only j, and not Δ. However, he does not obtain Theorem 3 in full, but shows instead that one has either Theorem 3 or $j(\underline{k}_{\underline{p}}.\underline{k}) \equiv j(\underline{k})^p \bmod \underline{p}$.

V CLASS FIELDS

(K. Iwasawa, Nov. 20 and 27, 1957)

§1. Introduction.

As was proved in IV, § 5, the class invariants $j(\underline{k})$ of an imaginary quadratic field K satisfy certain fundamental congruences. In the following, we shall first give an outline of class field theory in its classical form and then show, using that theory, how we can deduce from those congruences arithmetic properties of the numbers $j(\underline{k})$ and of the extension $K(j(\underline{k}))/K$. For the details of classical class field theory, cf. Hasse's Klassenkörper Bericht quoted in I, §4.

§2. Ideal groups.

Let K be a finite algebraic number field. Let m be a divisor of K, i.e. a formal product of a finite number of prime divisors of K : $m = \pi \underline{p}_i^{e_i}$. Put $m = m_0 m_\infty$, where m_0 and m_∞ are, respectively, products of non-archimedean and archimedean prime divisors in $m = \pi \underline{p}_i^{e_i}$; m_0 may be then identified with the corresponding ideal of K. In the following, we shall always consider such a divisor m that m_0 is an integral ideal of K and that m_∞ is a product of a number of distinct real archimedean prime divisors of K. For a number ξ in K, we then write

$$\xi \equiv 1 \mod. m$$

if and only if $\nu_{\underline{p}_i}(\xi - 1) \geqq e_i$ for every $\underline{p}_i$ dividing m_0 and $\sigma_{\underline{p}_j}(\xi) > 0$ for every $\underline{p}_j$ dividing m_∞; here $\nu_{\underline{p}_i}$ denotes the normalized exponential valuation of K belonging to $\underline{p}_i$ and $\sigma_{\underline{p}_j}$ denotes the isomorphism of K into the real field corresponding to $\underline{p}_j$.

Now, let $I = I(K)$ be the multiplicative group of all non-zero ideals of K. For any divisor m as considered above, we denote by $I_m = I_m(K)$ the subgroup of I consisting of all ideals of K which are prime to m_o and by $S_m = S_m(K)$ the group of all principal ideals (ξ) with $\xi \equiv 1$ mod. m. S_m is then a subgroup of I_m and I_m / S_m is a finite group. Any group H such that $S_m \subset H \subset I_m$ will be called an ideal group of K defined mod. m.

§3. The density of a set of prime ideals.

Let $P = P(K)$ be the set of all prime ideals of K. For any subset M of P, we put

$$\mu(s; M) = \sum_{\underline{p} \in M} N_{K/Q}(\underline{p})^{-s}$$

The right hand side is absolutely convergent for $R(s) > 1$ and $\mu(s; M)$ is well defined in that domain. Now, if the limit

$$\delta(M) = \lim_{\substack{s \to 1 \\ s > 1}} \frac{\mu(s; M)}{\mu(s; P)}$$

exists, we call $\delta(M)$ the density (Dirichlet density or Kronecker density) of the set M in P. Clearly, $\delta(P) = 1$.

Using the fact that the zeta-function $\zeta_K(s) = \prod_{\underline{p} \in P} (1 - N_{K/Q}(\underline{p})^{-s})^{-1}$ has a simple pole at $s = 1$, we can see by a simple computation that

$$\mu(s; P) \sim \log \zeta_K(s) \sim \log \frac{1}{s-1},$$

where $\sim$ indicates that the difference of the both sides of $\sim$ is a function bounded at $s = 1$. Hence

$$\lim_{s \to 1} \mu(s; P) / \log \frac{1}{s-1} = 1,$$

and the density $\delta(M)$ can be defined also by

$$\delta(M) = \lim_{s \to 1} \mu(s;\ M) \ / \ \log \frac{1}{s-1}.$$

Example. Let $N_{K/Q}(\underline{p}) = p^d$ (p = rational prime), $d = \deg \underline{p}$ = absolute degree of $\underline{p}$. Let $P' = P'(K)$ be the set of all $\underline{p}$ with $d = 1$ and let $P'' = P''(K)$ be the complement of P' in P. Since $N_{K/Q}(\underline{p}) = p^d \geqq p^2$ for $\underline{p}$ in P'', $\mu(s;\ P'')$ is bounded at $s = 1$. Hence, $\lim \mu(s;\ P'') \ / \ \log \frac{1}{s-1} = 0$, and

$$\delta(P'') = 0, \qquad \delta(P') = \delta(P) = 1 .$$

More generally, if we put $M' = M \cap P'$, $M'' = M \cap P''$ for any subset M of P, then

$$\delta(M'') = 0, \qquad \delta(M') = \delta(M),$$

whenever one of $\delta(M')$ and $\delta(M)$ exists.

Now, let H be an ideal group of K defined mod. m and let $\underline{k}$ be any class (coset) of I_m/H. For the set $\underline{k} \cap P$, we have then

(1) $$\delta(\underline{k} \cap P) = \delta(\underline{k} \cap P') = \frac{1}{h},$$

where $h = [I_m : H]$. This is the theorem of arithmetic progression for the field K, which generalizes the well-known theorem of Dirichlet for $K = Q$.

An outline of the proof of E. Hecke for (1) is as follows (cf. Göttinger Nachrichten, 1917): Let

$$L(s;\ \chi) = \pi'(1 - \chi(\underline{p})\ N_{K/Q}(\underline{p})^{-s})^{-1}, \qquad R(s) > 1,$$

be the L-function of K defined for a character χ of I_m/H (π' : over $\underline{p}$ in I_m). It is first proved that $L(s;\ \chi)$ can be analytically continued to a meromorphic function of s on the entire s-plane which has a unique pole of order 1 at $s = 1$ if $\chi = \chi_0$ and is holomorphic everywhere if $\chi \neq \chi_0$; here χ_0 denotes the principal character of I_m/H. On the other hand, it follows easily from the definition that

$$\pi_{\chi} L(s; \chi) = \pi'(1 - N_{K/Q}(\underline{p})^{-fs})^{\frac{h}{f}}, \qquad R(s) > 1,$$

where π_{χ} is taken over all characters of I_m/H and $f = f(\underline{p})$ is the order of the class of $\underline{p}$ in I_m/H. Hence,

$$\pi_{\chi} L(s; \chi) = \sum_{\nu=1}^{\infty} \frac{a_\nu}{\nu^s}, \qquad R(s) > 1,$$

with rational integers $a_\nu \geq 0$. Suppose now that $L(1; \chi) = 0$ for some $\chi \neq \chi_0$. Then the left hand side of the above would be holomorphic on the entire s-plane and, by a theorem of Landau on the convergence of Dirichlet series with positive coefficients, the right hand side of the above must converge for all s. Since a_ν are integers, it would then follow that $a_\nu = 0$ except for a finite number -- a contradiction. Hence:

$$L(1; \chi) \neq 0 \qquad \chi \neq \chi_0.$$

Now, by a simple computation, we have

$$\log L(s; \chi) = \Sigma' \sum_{\nu=1}^{\infty} \frac{1}{\nu} \chi(\underline{p})^\nu N_{K/Q}(\underline{p})^{-\nu s} \sim \Sigma' \chi(\underline{p}) N_{K/Q}(\underline{p})^{-s} \qquad s > 1.$$

By the orthogonality of the characters χ, it then follows that

$$\sum_{\chi} \chi(\underline{k}^{-1}) \log L(s; \chi) \sim h \cdot \sum_{\underline{p} \in \underline{k}} N_{K/Q}(\underline{p})^{-s} = h \cdot \mu(s; \underline{k} \cap P),$$

for any class $\underline{k}$ of I_m/H. Since $L(1; \chi) \neq 0$ for $\chi \neq \chi_0$, we have

$$\log L(s; \chi_0) \sim \log \frac{1}{s-1}; \quad \log L(s; \chi) \sim 0, \qquad \chi \neq \chi_0,$$

and the left hand side of the above is $\sim \log \frac{1}{s-1}$. Hence we get

$$\delta(\underline{k} \cap P) = \lim_{s \to 1} \mu(s; \underline{k} \cap P) / \log \frac{1}{s-1} = \frac{1}{h}, \qquad \text{q.e.d.}$$

Remark. By Theorems II, IV of §6 below (which can be proved purely algebraically), there exists an abelian extension E of degree h over K such that the zeta-function $\zeta_E(s)$ of E is the product of $\pi_{\chi} L(s; \chi)$ and a finite number of factors of the form $(1 - q^{-s})^{-1}$, $q > 1$. Since $\zeta_E(s)$ has a simple pole at $s = 1$, this immediately implies $L(1; \chi) \neq 0$ for $\chi \neq \chi_0$. For K = Q, such an E is given by a subfield of a cyclotomic field.

§4. The inequality $h \leq n$.

Let L be a Galois extension of degree n over K. A prime ideal $\underline{p}$ of K is said to be completely decomposed in L if $\underline{p}$ is decomposed into the product of n distinct prime ideals of L, or equivalently, if $\underline{p}$ is unramified in L and $\underline{p} = N_{L/K}(\underline{P})$ for some prime ideal $\underline{P}$ of L. Let W denote the set of all such $\underline{p}$ of K which are completely decomposed in L, and let $P'''(L)$ be the set of all prime ideals $\underline{P}$ of L such that $\deg \underline{P} = 1$ and that $\underline{p} = N_{L/K}(\underline{P})$ is unramified in L. Then $\underline{P} \longrightarrow \underline{p} = N_{L/K}(\underline{P})$ defines an n to 1 correspondence between the sets $P'''(L)$ and $W' = W \cap P'(K)$ and, as $N_{L/Q}(\underline{P}) = N_{K/Q}(\underline{p})$, we get

$$\mu(s; P'''(L)) = n\mu(s; W').$$

However, the difference of $P'(L)$ and $P'''(L)$ is a finite set and $\delta(P'''(L)) = \delta(P'(L)) = \delta(P(L)) = 1$. Hence, it follows from the above that

(2) $$\delta(W) = \delta(W') = \frac{1}{n}.$$

Now, let m be a divisor of K as considered in §2. Define a group $N_m = N_m(L/K)$ by

$$N_m = S_m(K)\, N_{L/K}(I_m(L)).$$

Clearly, $S_m \subset N_m \subset I_m$ and we put

$$h_m = h_m(L/K) = [I_m : N_m].$$

By the definition of W, $W_m = W \cap I_m$ is contained in $N_{L/K}(I_m(L))$ and, hence, also in N_m. As the difference of W and W_m is again a finite set, we obtain from (1) and (2) that

$$\frac{1}{n} = \delta(W) = \delta(W_m) \leq \delta(N_m \cap P) = \frac{1}{h_m},$$

namely that

(3) $$h_m \leq n, \quad \text{or} \quad [I_m : N_m(L/K)] \leq [L : K].$$

This is called the <u>second fundamental inequality</u> of class field theory and it is valid for any m and for any finite Galois extension L of K.

§5. Definition of class fields.

Now, a finite Galois extension L of a finite algebraic number field K is called a class field over K if, in the above, $h_m = n$ holds for some divisor m of K. If L/K is such a class field, there exists a unique divisor f of K such that $h_m = n$ holds if and only if m is a multiple of f. The divisor $f = f(L/K)$ is then called the conductor of L/K.

Criterion. If L/K is a class field, there exists a divisor m of K and an ideal group H of K defined mod. m such that

i) $N_{L/K}(I_m(L)) \subset H$,

ii) $H \cap P$ is contained in W except for a subset of density 0.

Conversely, if there exist such m and H, then L/K is a class field and

$$h_m = n, \qquad H = N_m.$$

Proof. If L/K is a class field, put $H = N_m$ for a divisor m such that $h_m = n$. Then i) holds obviously and ii) follows from that the density of the difference of $N_m \cap P$ and W_m is 0. Suppose, conversely, such m and H exist. By i), we have $S_m \subset N_m \subset H \subset I_m$. Put $a = [H : N_m]$. Then, by the theorem of arithmetic progression, $\delta(H \cap P) = \frac{a}{h_m}$. However, by ii), $\delta(H \cap P) \leq \delta(W) = \frac{1}{n}$. Hence $\frac{a}{h_m} \leq \frac{1}{n}$, or, $an \leq h_m$. It then follows from (3) that $a = 1$, $h_m = n$, q.e.d.

As an immediate application we notice the following: Let both L_1/K and L_2/K be class fields and let m be divisible by $f(L_1/K)$ and $f(L_2/K)$. It is then easy to see that m and $H = N_m(L_1/K) \cap N_m(L_2/K)$ satisfy the above conditions i) and ii) for the extension $L_1 L_2/K$. Hence $L_1 L_2/K$ is also a class field and $h_m(L_1 L_2/K) = [L_1 L_2 : K]$, $N_m(L_1 L_2/K) = N_m(L_1/K) \cap N_m(L_2/K)$. Now, the condition $N_m(L_1/K) \subset N_m(L_2/K)$ is obviously equivalent with the condition $N_m(L_1 L_2/K) = N_m(L_1/K)$ and, hence, also with $[L_1 L_2 : K] = h_m(L_1 L_2/K) = h_m(L_1/K) = [L_1 : K]$, or, $L_1 L_2 = L_1$. Therefore, $L_1 \supset L_2$ if and only if $N_m(L_1/K) \subset N_m(L_2/K)$.

§6. Fundamental theorems.

We are now going to state the fundamental theorems of class field theory.

I. A finite extension L of K is a class field over K if and only if L/K is an abelian extension (i.e. a Galois extension with an abelian Galois group).

The most essential step of the proof is to prove

$$n \leqq h_m$$

for some m, when L/K is a cyclic extension. This is called the first fundamental inequality of class field theory.

II. Given any ideal group H of K defined mod. m, there exists a unique class field L over K such that $[I_m : N_m(L/K)] = [L : K]$ and $H = N_m(L/K)$.

III. A prime divisor p of K is ramified in a class field L over K if any only if p divides the conductor f(L/K) of L/K.

Before going to state the next theorem, Artin's reciprocity law, we first make some preparations. Let L/K be an arbitrary Galois extension of degree n with the Galois group G. Let $\underline{p}$ be a prime ideal of K unramified in L, and $\underline{P}$ a prime ideal of L dividing $\underline{p}$. Then there exists a unique element σ in G such that

$$\sigma(\alpha) \equiv \alpha^{N_{K/Q}(\underline{p})} \mod. \underline{P} \tag{4}$$

for any algebraic integer α in L. This element σ is called the Frobenius substitution of $\underline{P}$ for the extension L/K and is denoted by $\sigma_{\underline{P}}$; it generates the so-called decomposition group of the prime ideal $\underline{P}$, and if f is the order of $\sigma_{\underline{P}}$ and n = fg, then

$$N_{L/K}(\underline{P}) = \underline{p}^f, \qquad \underline{p} = \underline{P}_1, \dots, \underline{P}_g, \qquad \underline{P}_1 = \underline{P},$$

where $\underline{P}_i$ are distinct prime ideals of L.

Now, assume that L/K is a class field, i.e. an abelian extension. Then the Frobenius substitutions of $\underline{P}_1, \ldots \underline{P}_g$ coincide with each other and we may denote these $\sigma_{\underline{P}_i}$ $(i = 1, \ldots g)$ simply by $\sigma_{\underline{p}}$. Let m be a divisor of K such that $h_m = n$ and let $\mathfrak{a} = \prod_{i=1}^{r} \underline{p}_i^{e_i}$ be the prime ideal decomposition of an ideal $\mathfrak{a}$ in I_m. By III, every $\underline{p}_i$ is unramified in L and $\sigma_{\underline{p}_i}$ is defined as above. So, we may define $\sigma_{\mathfrak{a}}$ by

$$\sigma_{\mathfrak{a}} = \prod_{i=1}^{r} \sigma_{\underline{p}_i}^{e_i} .$$

Clearly, $\mathfrak{a} \longrightarrow \sigma_{\mathfrak{a}}$ is a homomorphism of I_m into G. But we have the following theorem:

IV. <u>The homomorphism</u> $\mathfrak{a} \longrightarrow \sigma_{\mathfrak{a}}$ <u>induces an isomorphism</u>

$$I_m / N_m \cong G;$$

<u>namely, the homomorphism is surjective and the kernel is</u> N_m.

An immediate consequence of IV is the following: Prime ideals of K, which are contained in the same class of I_m / N_m, are decomposed into the product of the same number of prime factors in L. In particular, every prime ideal in N_m is completely decomposed in L without exception.

Finally, let L/K be again a Galois extension of degree n and let σ be an arbitrary element of the Galois group G of L/K. We denote by P_σ the set of all prime ideals $\underline{p}$ of K such that $\underline{p}$ is unramified in L and that $\sigma = \sigma_{\underline{P}}$ for some prime ideal $\underline{P}$ of L dividing $\underline{p}$. Then

V. <u>If</u> c <u>is the number of elements in the conjugate class of</u> σ <u>in</u> G, <u>then</u>

$$\delta(P_\sigma) = \frac{c}{n}.$$

Notice that if L/K is a class field, V is an immediate consequence of IV and the theorem of arithmetic progression.

§7. Hilbert's absolute class field.

Let 1 denote the unit divisor of K. Then $I_1(K) = I(K)$, and $S_1(K)$ is the group of all principal ideals of K. The factor group I_1/S_1 is hence the ideal class group C_K of K in the ordinary sense and its order h is nothing but the class number of K.

Now, by the existence theorem II, there exists a class field L_0 over K such that

$$h = [I_1 : S_1] = [L_0 : K], \qquad S_1 = N_1\,(L_0/K).$$

The conductor $f(L_0/K)$ is then 1 and, by III, L_0/K is an unramified extension. Take an arbitrary unramified abelian extension L/K. Again by III, the conductor $f(L/K)$ is 1, and since $S_1 = N_1(L_0/K)$ is contained in $N_1(L/K)$, L is contained in L_0 (cf. the last lines of §5). Thus we see that L_0 is the unique maximal unramified abelian extension of K. L_0 is called the Hilbert's absolute class field over K. By IV, the Galois group of L_0/K is canonically isomorphic with the ideal class group $C_K = I_1/S_1$ of K.

§8. Conjugacy of class invariants.

We now assume that K is an imaginary quadratic field and denote by $\underline{k}_1, \dots, \underline{k}_h$ the ideal classes of $C_K = I_1/S_1$. The class invariants $j(\underline{k}_i)$, $i = 1, \dots, h$, are then distinct algebraic integers. Take a finite Galois extension L over K containing all such class invariants $j(\underline{k}_i)$ and denote by $\xi_1, \dots, \xi_s$ the distinct elements in the set $\sigma(j(\underline{k}_i))$ where σ runs over the Galois group G of L/K and $1 \leq i \leq h$. We then put

$$\eta = \prod_{a<b} (\xi_a - \xi_b)$$

and denote by P^* the set of all prime ideals $\underline{p}$ of K with $\deg \underline{p} = 1$ such that $\underline{p}$ is unramified in L and is prime to η. It is clear that

$\delta(P^*) = \delta(P') = 1$. Hence, for any subset M of P(K) with $\delta(M) > 0$, $\delta(M \cap P^*) = \delta(M) > 0$ and the intersection $M \cap P^*$ is non-empty.

Let $\underline{p}$ be a prime ideal in P^*, $\underline{k}_{\underline{p}}$ the ideal class of C_K containing $\underline{p}$, and $p = N_{K/Q}(\underline{p})$. By IV, §5 , we have then

$$j(\underline{k}_{\underline{p}}^{-1}\underline{k}) \equiv j(\underline{k})^p \quad \text{mod. } \underline{p},$$

for any ideal class $\underline{k}$. On the other hand, if $\underline{P}$ is a prime ideal of L dividing $\underline{p}$ and $\sigma_{\underline{P}}$ is the Frobenius substitution of $\underline{P}$ for L/K, then by (4),

$$\sigma_{\underline{P}}(j(\underline{k})) \equiv j(\underline{k})^p \quad \text{mod.} \underline{P}.$$

It follows that

$$\sigma_{\underline{P}}(j(\underline{k})) \equiv j(\underline{k}_{\underline{p}}^{-1}\underline{k}) \quad \text{mod. } \underline{P}.$$

But, the both sides of the above congruence are numbers in the set $\xi_1, \ldots, \xi_s$. Hence, as $\underline{p}$ is prime to η , we get

(5) $$\sigma_{\underline{P}}(j(\underline{k})) = j(\underline{k}_{\underline{p}}^{-1}\underline{k}), \qquad \underline{p} \in P^*, \ \underline{P} \mid \underline{p}.$$

Now, let $\underline{k}$ and $\underline{k}'$ be arbitrary ideal classes of C_K. By the theorem of arithmetic progression, $\delta(\underline{k}\,\underline{k}'^{-1} \cap P^*) = \delta(\underline{k}\,\underline{k}'^{-1} \cap P) > 0$ and the ideal class $\underline{k}\,\underline{k}'^{-1}$ contains a prime ideal $\underline{p}$ in P^* so that $\underline{k}' = \underline{k}_{\underline{p}}^{-1}\underline{k}$. It then follows from (5) that the class invariants $j(\underline{k})$ and $j(\underline{k}')$ are conjugate in L over K. On the other hand, given any σ in G, $\delta(P_\sigma \cap P^*) = \delta(P_\sigma) > 0$ by V and there exists a prime ideal $\underline{p}$ in P^* such that $\sigma = \sigma_{\underline{P}}$ for some $\underline{P}$ dividing $\underline{p}$. Hence, by (5), any conjugate $\sigma(j(\underline{k}))$ of a class invariant $j(\underline{k})$ is another class invariant $j(\underline{k}')$. Thus:

Theorem 1. _The class invariants_ $j(\underline{k}_i)$, $i = 1, \ldots, h$, _of an imaginary quadratic field_ K _form a complete set of conjugates over_ K.

§9. _The extension_ $K(j(\underline{k})) / K$.

From now on, we put $L = K(j(\underline{k}_1), \ldots, j(\underline{k}_h))$, for the latter is a Galois extension of K by Theorem 1.

Lemma. *A prime ideal $\underline{p}$ in P^* is completely decomposed in L if and only if $\underline{p}$ is contained in S_1* (i.e. a principal ideal).

Proof. Let $\underline{P}$ be a prime ideal of L dividing $\underline{p}$. Then $\underline{p}$ is completely decomposed in L if and only if $\sigma_{\underline{P}} = 1$. Since $L = K(j(\underline{k}_1), \ldots, j(\underline{k}_h))$, it follows from (5) that $\sigma_{\underline{P}} = 1$ if and only if $j(\underline{k}_{\underline{p}}^{-1}\underline{k}) = j(\underline{k})$ for every $\underline{k}$, namely, if and only if $\underline{k}_{\underline{p}} = 1$, or, $\underline{p} \in S_1$.

We now verify that the two conditions of the criterion given in §5 are satisfied for $m = 1$, $H = S_1$ and for the extension L/K. Let $\mathfrak{A}$ be an arbitrary ideal of L. By the theorem of arithmetic progression, there exists a prime ideal $\underline{P}$ of L such that $\deg \underline{P} = 1$ and $\mathfrak{A} = (\alpha)\underline{P}$ with a number α in L. We may also assume that $\underline{p} = N_{L/K}(\underline{P})$ is prime to $\mathfrak{m}$. Then $\underline{p}$ is contained in P^* and is completely decomposed in L. By the above lemma, $\underline{p}$ is hence contained in S_1, and so is the ideal $N_{L/K}(\mathfrak{A}) = N_{L/K}(\alpha)\, N_{L/K}(\underline{P})$. Thus, i) is verified. As $\delta(S_1 \cap P^*) = \delta(S_1 \cap P)$, the condition ii) is also satisfied by the above lemma.

By the criterion, L/K is therefore a class field and

$$h = [L : K], \qquad S_1 = N_1(L/K).$$

As explained in §7, L is hence the Hilbert's absolute class field over K, i.e. the maximal unramified abelian extension of K. Since L/K is now abelian and $j(\underline{k}_i)$, $1 \leq i \leq h$, are conjugate over K, we can obtain L by adjoining just one $j(\underline{k}_i)$ to K. Thus the following theorem is proved:

Theorem 2. *Let $j(\underline{k})$ be any class invariant of an imaginary quadratic field* K. *Then $K(j(\underline{k}))$ is the maximal unramified abelian extension of* K.

As mentioned in §7, we have then a canonical isomorphism between the ideal class group C_K of K and the Galois group G of L/K:

$$C_K \cong G.$$

Let $\sigma_{\underline{k}}$ denote the element of G corresponding to an ideal class $\underline{k}$ under this isomorphism. If $\underline{p}$ is a prime ideal of P^* contained in the class $\underline{k}$, then $\sigma_{\underline{k}} = \sigma_{\underline{p}}$ by the definition of the isomorphism. We get therefore from (5) the following explicit formula for the automorphism $\sigma_{\underline{k}}$ of L/K:

(6) $$\sigma_{\underline{k}}(j(\underline{k}')) = j(\underline{k}^{-1}\underline{k}'), \qquad \underline{k}, \underline{k}' \in C_K.$$

Now, let τ be an automorphism of the field of all algebraic numbers. Clearly, $\tau(L)$ is the maximal unramified abelian extension of $\tau(K)$. Since K is a quadratic field over Q and $\tau(K) = K$, it follows that $\tau(L) = L$. Therefore, L/Q is also a Galois extension. Let $\mathcal{G}$ denote the Galois group of L/Q and let ρ be the element of $\mathcal{G}$ which maps α in L to $\bar{\alpha}$ (conjugate complex of α). Clearly, $\rho^2 = 1$. But ρ induces a non-trivial automorphism on the imaginary field K. Hence the order of ρ is 2 and $\mathcal{G}$ is the semi-direct product of the subgroup $\{1, \rho\}$ and the normal subgroup G. Let $\mathfrak{a}$ be an ideal in an ideal class $\underline{k}$. Then $\rho(j(\underline{k})) = \overline{j(\underline{k})} = \overline{j(\mathfrak{a})} = j(\bar{\mathfrak{a}})$. Since $\mathfrak{a}\bar{\mathfrak{a}} = N_{K/Q}(\mathfrak{a})$ is a principal ideal, $\bar{\mathfrak{a}}$ is in the class $\underline{k}^{-1}$ and we have

(7) $$\rho(j(\underline{k})) = j(\underline{k}^{-1}).$$

Combined with Theorem 1, this gives us the following

Theorem 1'. *The class invariants* $j(\underline{k}_i)$, $i = 1, \ldots, h$, *of an imaginary quadratic field* K *form a complete set of conjugates over* Q.

Therefore, if we put $f(t) = \prod_{i=1}^{h} (t - j(\underline{k}_i))$, f(t) is an irreducible polynomial of t with integral rational coefficients.

Another consequence of (7) is the following: Let σ be an arbitrary element of G and let $\sigma = \sigma_{\underline{k}}$, $\sigma^{-1} = \sigma_{\underline{k}^{-1}}$. Since G is normal in $\mathcal{G}$, $\rho\sigma\rho^{-1}$ is again contained in G, and it follows from (6) and (7) that

$$(\rho\sigma\rho^{-1})(j(\underline{k}')) = (\rho\sigma)(j(\underline{k}'^{-1})) = \rho(j(\underline{k}^{-1}\underline{k}'^{-1})) = j(\underline{kk}') = \sigma^{-1}(j(\underline{k}')),$$

for any class invariant $j(\underline{k}')$. Hence

$$\rho\sigma\rho^{-1} = \sigma^{-1}, \qquad \sigma \in G.$$

This, together with the fact that $\mathcal{G}$ is the semi-direct product of $\{1, \rho\}$ and G, shows that the structure of the Galois group $\mathcal{G}$ is completely determined when we know the structure of $G \cong C_K$.

VI REMARKS ON CLASS-INVARIANTS AND RELATED TOPICS

(S. Chowla, Dec. 3 and Dec. 10, 1957)

§1. Let $p > 3$, be a prime. Suppose that the class-number of the imaginary quadratic field $R(\sqrt{-p})$ is 1. In this case we have seen that

(1) $$j\left(\frac{-3+\sqrt{-p}}{2}\right)$$

is a rational integer. Here

(2) $$j(\omega) = \frac{\left\{1 + 240 \sum_1^\infty \sigma_3(n) q^{2n}\right\}^3}{q^2 \prod_1^\infty (1-q^{2n})^{24}}$$

and $q = e^{\pi i \omega}$. It is a remarkable fact (see Weber, Lehrbuch der Algebra, Bd. 3, 457-462) that not only the number (i) but also its cube-root is a rational integer. Under the same restriction on the class-number it is also true (Weber, p. 504) that

$$\frac{\sqrt{j(\omega) - 1728}}{\sqrt{-p}} \qquad \left\{\omega = \frac{-1+\sqrt{-p}}{2}\right\}$$

is a rational integer.

We can now formulate the

Theorem If the class-number of $R(\sqrt{-p})$ is 1, where p is a prime > 3, then the diophantine equation

$$x^3 - py^2 = -1728$$

has a solution in rational integers with $x = \left\{e^{\pi\sqrt{p}/3}\right\}$ where $\{u\}$ denotes the integer nearest to u.

It is believed that $p = 7, 11, 19, 43, 67, 163$ are the only primes > 3 such that the class-number of $R(\sqrt{-p})$ is 1. Heilbronn and Linfoot have proved that there is at most one more prime p with the property in question (by D. H. Lehmer we must have $p > 10^9$ for such a prime). Chowla and Selberg (Proc. Nat. Acad. Sci. U.S.A. 1949) have proved that the existence of a prime $p > 163$ with class-number of $R(\sqrt{-p})$ equal to 1, would imply the falseness of the "extended Riemann hypothesis". The formulation of the above theorem is not without interest for the following reason. Siegel's method (Acta Arithmetica, Bd. 1, 1936) does not lead to a determinable constant c beyond which the class-number of $R(\sqrt{-p})$ is greater than 1. Also the Thue-Siegel-Roth theorem on the finiteness of the number of solutions of certain diophantine equations (including the form $ay^2 = bx^3 + f$ of our theorem) does not lead to determinable constants c such that the equations have no solution when the variables in the diophantine equation are absolutely greater than c. Our theorem would be useful if, for example, one could use the methods of Delaunay and Nagell to prove the existence of determinable constants $c_1 (< \frac{\pi}{3})$ and c_2 such that $x^3 - py^2 = -1728$ has no solutions in integers for $x > e^{c_1\sqrt{p}}$ and $p > c_2$. One could then deduce from our theorem that there is a determinable constant c_3 such that the class-number of $R(\sqrt{-p})$ is greater than 1 for $p > c_3$. It should also be remarked here that a recent claim by Kurt Heegner, (M.Z. 1954) to have solved this problem, seems unjustified.

§ 2. If m is a positive integer such that the classes of reduced binary quadratic forms of discriminant $-m$ have a single class in each genus, one uses Kronecker's "Grenz Formel" to determine explicitly (i) the values of $j(\omega)$ when $\omega = i\sqrt{m}$.

We introduce the functions

$$\eta(\omega) = q^{\frac{1}{12}} \prod_1^\infty (1 - q^{2n})$$

$$f(\omega) = e^{\frac{-\pi i}{24}} \frac{\eta(\frac{\omega+1}{2})}{\eta(\omega)}$$

(see Weber, p. 113)

Kronecker's formula ("Grenz Formel") states that (ibid, p. 531)

$$\lim_{S \to 1+0} \left\{ \Sigma' \, (Ax^2 + 2Bxy + Cy^2)^{-S} - \frac{\pi}{(S-1)\sqrt{m}} \right\}$$

$$= - \frac{2\pi \Gamma'(1)}{\sqrt{m}} + \frac{\pi}{\sqrt{m}} \log \frac{A}{4m}$$

$$- \frac{2\pi}{\sqrt{m}} \log \eta(\omega_1)\, \eta(\omega_2)$$

where $m = AC - B^2$ and

$$\omega_1 = \frac{B + i\sqrt{m}}{A}, \quad \omega_2 = \frac{-B + i\sqrt{m}}{A}$$

(The prime in Σ' means that $x = y = 0$ is omitted from the summation where x,y range over all integers.) From this, without difficulty, one deduces, for example, that

$$\left(\frac{f(\sqrt{-105})}{\sqrt[4]{2}}\right)^{12} = (2 + \sqrt{5})^2 \, (55 + 12\sqrt{21}) \, (6 + \sqrt{35}) \, (2 + \sqrt{3})^3$$

Using the formula

$$j(\omega) = \frac{\{f^{24}(\omega) - 16\}^3}{f^{24}(\omega)}$$

we can now deduce the value in algebraic form of $j(\sqrt{-105})$.

§3. We add a few remarks on Ramanujan's function $\tau(n)$ defined by

(4)
$$\sum_1^\infty \tau(n)\, x^n = x \prod_1^\infty (1 - x^n)^{24}$$

$$(|x| < 1)$$

[cf. the denominator of (2)]. Ramanujan conjectured and Mordell proved that the function $\tau(n)$ is multiplicative, i.e.

(5) $$\tau(mn) = \tau(m)\ \tau(n) \text{ if } (m,n) = 1$$

Ramanujan also conjectured that for prime p,

(6) $$|\tau(p)| < 2p^{\frac{11}{2}}$$

the best that is known in this direction is that $\tau(p) = O(p^{23/4})$ which follows from A. Weil's estimate on Kloosterman's sums

(7) $$\left| \sum_{x=1}^{p-1} e^{\frac{2\pi i}{p}(x+\bar{x})} \right| < 2\sqrt{p}$$

Here $\bar{x}$ is defined by $x\bar{x} \equiv 1 \pmod p$. Another unsolved problem (raised by D. H. Lehmer) is whether $\tau(n)$ can ever be 0.

§4. The calculation of class-invariants $j(\omega)$ can also be made to depend on the classical modular equations of Legendre and Jacobi. We recall the main details of the connection. For $0 < k < 1$ write

(8) $$K = \int_0^{\frac{\pi}{2}} \frac{d\phi}{\sqrt{1-k^2\sin^2\phi}} = \frac{\pi}{2} F(\tfrac{1}{2}, \tfrac{1}{2}, 1; k^2)$$

k' is defined by $k^2 + k'^2 = 1$ and

(9) $$K' = \int_0^{\frac{\pi}{2}} \frac{d\phi}{\sqrt{1-k'^2\sin^2\phi}}$$

Also write

$$q = \exp(-\pi K'/K)$$

Then

(10) $$\sqrt{\frac{2K}{\pi}} = \sum_{-\infty}^{\infty} q^{n^2}$$

Let

$$L = \int_0^{\frac{\pi}{2}} \frac{d\phi}{\sqrt{1-\ell^2\sin^2\phi}} \quad , \quad L' = \int_0^{\frac{\pi}{2}} \frac{d\phi}{\sqrt{1-\ell'^2\sin^2\phi}}$$

If

$$\frac{L'}{L} = \frac{nK'}{K}$$

where n is a positive integer there is an algebraic relation between k and ℓ called the "modular equation"):

n = 3: $\sqrt{k\ell} + \sqrt{k'\ell'} = 1$

n = 7 $\sqrt[4]{k\ell} + \sqrt[4]{k'\ell'} = 1$

Both results are due to Legendre. By setting $k = \ell'$ (and so $k' = \ell$) we get an algebraic equation for k when $\frac{K'}{K} = \sqrt{n}$ and n is a positive integer. Abel remarked that the resulting equation for k is "solvable by radicals".

We quote the following results

(11) If $\frac{K'}{K} = \sqrt{210}$, then $k = (\sqrt{2} - 1)^4(2 - \sqrt{3})^2(\sqrt{7} - \sqrt{6})^4$

$$\times(8 - 3\sqrt{7})^2(\sqrt{10} - 3)^4(4 - \sqrt{15})^2(\sqrt{15} - \sqrt{14})^2$$

$$\times(6 - \sqrt{35})^2$$

(Ramanujan in a letter to Hardy dated 27 Feb. 1913)

(12) If $\frac{K'}{K} = \sqrt{47}$, then $\sqrt[12]{16\,kk'}$ is the real root of

(13) $$x^5 + 2x^4 + 2x^3 + x^2 - 1 = 0$$

(after multiplication by x - 1 the left-side becomes $x^6 + x^5 - x^3 - x^2 - x + 1$)

The following solution originated from G. P. Young (Am. J. of Math. 10, 1888, 99-130) and was simplified by Cayley (ibid, 13, 1891, 53-58). The real root of (13) is

$$\sqrt{\frac{\sqrt[5]{A} + \sqrt[5]{B} + \sqrt[5]{C} + \sqrt[5]{D}}{10}}$$

where

$$A, D = 39000 + 18200\sqrt{5} \pm (1720 + 920\sqrt{5})\sqrt{235 + 94\sqrt{5}}$$

$$B, C = 39000 - 18200\sqrt{5} \mp (1720 - 920\sqrt{5})\sqrt{235 - 94\sqrt{5}}$$

(See also Journal of the Indian Math. Soc. 18, 1929-30, p. 273 of "Notes and questions").

§5. Finally we quote the following results of Deuring ["Die Typen der Multiplikatorenringe elliptischer Funktionenkörper"] (G. Herglotz zum 60. Geburtstag gewidmet) in Abh. aus der Math. Sem. der Hansischen Univ. 14, 1941, 197-272.

Let

$$j(\omega) = \frac{(f^{24}(\omega) - 16)^3}{f^{24}(\omega)}$$

where $\omega = \omega_1/\omega_2$ is the quotient of two numbers ω_1 and ω_2 which form a basis of any ideal in an imaginary quad. field Σ then by Weber (ibid, 540-541)

$$f(\omega)/\sqrt{2}$$

is a unit of an algebraic number field if Q is "voll zerlegt" in Σ, otherwise the product of $\frac{f(\omega)}{\sqrt{2}}$ by a certain power of 2 with positive exponent is a unit. From this result of Weber (based on a simple argument using Kronecker's Grenz Formel), Deuring deduces:

A singular invariant j of characteristic 0 is either not divisible by any prime factor of 2 or by every such factor according as 2 is "voll zerlegt" or not in Σ.

By more difficult arguments Deuring proves analogous results when 2 is replaced by 3, 5, 7 or 13. We quote from p. 271 of his paper (also p. 257 for a relevant table):

j ist durch keinen Primfaktor von 3 oder 5 teilbar oder durch jeden, je nachdem 3(5) in Σ voll zerlegt oder nicht.

$j - 2^6 \cdot 3^3 = j + 1$ ist durch keinen Primfaktor von 7 teilbar oder durch jeden, je nachdem 7 in Σ voll zerlegt oder nicht.

$j - 5$ ist durch keinen Primfaktor von 13 teilbar oder durch jeden, je nachdem 13 in Σ voll zerfällt oder nicht.

§6. In this section we shall give a new proof of the fundamental fact that the function

$$j(\tau) = \frac{\{1 + 240 \sum_{1}^{\infty} \sigma_3(n) e^{2n\pi i\tau}\}^3}{e^{2\pi i\tau} \prod_{1}^{\infty} (1 - e^{2n\pi i\tau})^{24}}$$

is invariant for the transformations $\tau \longrightarrow \tau + 1$ and $\tau \longrightarrow -\frac{1}{\tau}$ (the first is, of course, obvious). Here $\mathrm{Im}(\tau) > 0$ and

$$\sigma_a(n) = \sum_{d|n} d^a .$$

The proof is based on the functional equation for the Riemann-Zeta-Function

(1) $$2\Gamma(s)\zeta(s)\cos\frac{s\pi}{2} = (2\pi)^s \zeta(1 - s)$$

and on Mellin's Integral ($s = \sigma + it$)

(2) $$\frac{1}{2\pi i} \int_{(\sigma = k > 0)} \Gamma(s) y^{-s} ds = e^{-y} .$$

Here $\mathrm{Re}(y) > 0$ and the path of integration is the vertical line $\sigma = k$ where $k > 0$. We first prove the

Theorem Write

(3) $$H_a(y) = \sum_{1}^{\infty} \sigma_a(n) e^{-2n\pi y} + (-1)^{\frac{a+1}{2}} \frac{\Gamma(a+1)\zeta(a+1)}{(2\pi)^{a+1}} .$$

Then if a is an odd integer ≥ 3, we have

(4) $$(-1)^{\frac{a+1}{2}} y^{a+1} H_a(y) = H_a\left(\frac{1}{y}\right) .$$

There is a slightly different formulation for the case $a = 1$, which enables us to study the effect of the transformation $\tau \longrightarrow -\frac{1}{\tau}$ on the denominator of $j(\tau)$. For the numerator we use our theorem for the case $a = 3$. Combining these two results we deduce the property of $j(\tau)$ in question. Write

$$G(y) = G_a(y) = \sum_1^\infty \sigma_a(n) e^{-2n\pi y} \tag{5}$$

and

$$F(s) = \zeta(s)\zeta(s-a) \left[= \sum_1^\infty \frac{\sigma_a(n)}{n^s} \text{ for } \sigma > a+1 \right]. \tag{6}$$

From (2) and (6)

$$G(y) = \frac{1}{2\pi i} \int_{\sigma = a+2} \Gamma(s) F(s) (2\pi y)^{-s} ds . \tag{7}$$

From (1)

$$2\Gamma(s-a)\zeta(s-a) \cos \frac{(s-a)\pi}{2} = (2\pi)^{s-a} \zeta(1-s+a) . \tag{8}$$

Since a is odd it follows from (6), (1), and (8) that

$$2\Gamma(s)\Gamma(s-a)F(s)\sin(s\pi)(-1)^{\frac{a-1}{2}} = (2\pi)^{2s-a} F(a+1-s) . \tag{9}$$

Multiplying both sides by $\Gamma(a+1-s)$ and using $\Gamma(x)\Gamma(1-x) = \pi/\sin(\pi x)$ we get

$$\Gamma(s)F(s)(-1)^{\frac{a+1}{2}} = (2\pi)^{2s-a-1} \Gamma(a+1-s)F(a+1-s) . \tag{10}$$

From (7) and (10)

$$G(y) = \frac{1}{2\pi i} \int_{\sigma = a+2} (2\pi)^{2s-a-1} (2\pi y)^{-s} (-1)^{\frac{a+1}{2}} \Gamma(a+1-s) F(a+1-s) ds$$

$$= \frac{1}{2\pi i} \int_{t=-\infty}^{t=+\infty} (2\pi)^{a+3+2ti} (2\pi y)^{-a-2-ti} (-1)^{\frac{a+1}{2}} \Gamma(-1-ti) F(-1-ti) dt$$

(and now transform by $t \longrightarrow -t$)

$$= \frac{1}{2\pi i} \int_{t=-\infty}^{t=+\infty} (2\pi)^{a+3-2ti} (2\pi y)^{-a-2+ti} (-1)^{\frac{a+1}{2}} \Gamma(-1+ti) F(-1+ti) dt$$

$$= \frac{1}{2\pi i} \int_{\sigma=-1} (2\pi)^{a+1-2s} (2\pi y)^{-a-1+s} (-1)^{\frac{a+1}{2}} \Gamma(s) F(s) ds$$

so that

(11) $$G(y) = (-1)^{\frac{a+1}{2}} \left(\frac{1}{y}\right)^{a+1} \int_{\sigma=-1} \left(\frac{y}{2\pi}\right)^{s} \Gamma(s) F(s) ds \ .$$

Next, from (7),

$$G\left(\frac{1}{y}\right) = \frac{1}{2\pi i} \int_{(\sigma = a+2)} \left(\frac{y}{2\pi}\right)^{s} \Gamma(s) F(s) ds$$

$$= \frac{1}{2\pi i} \int_{(\sigma=-1)} \left(\frac{y}{2\pi}\right)^{s} \Gamma(s) F(s) ds$$

$$+ \left\{ \text{sum of residues of integrand at poles } s = a+1,\ s = 1,\ s = 0 \right\} \ .$$

Thus, using (11),

(12) $$G\left(\frac{1}{y}\right) = (-1)^{\frac{a+1}{2}} y^{a+1} G(y)$$

$$+ \left\{ \Gamma(a+1) \zeta(a+1) \left(\frac{y}{2\pi}\right)^{a+1} \right.$$

$$+ \Gamma(1) \zeta(1-a) \left(\frac{y}{2\pi}\right)$$

$$\left. + \zeta(0) \zeta(-a) \right\} \ .$$

Since a is odd and ≥ 3 it follows from (1) that $\zeta(1-a) = 0$, and so we have from the last relation:

$$G(\tfrac{1}{y}) = (-1)^{\frac{a+1}{2}} y^{a+1} G(y)$$

$$+ \Gamma(a+1)\,\zeta(a+1)\,(\tfrac{y}{2\pi})^{a+1}$$

$$+ \zeta(0)\,\zeta(-a) .$$

Using (1) and $\zeta(0) = -\frac{1}{2}$ this gives for odd $a \geq 3$,

$$(-1)^{\frac{a+1}{2}} y^{a+1} \left\{ G_a(y) + (-1)^{\frac{a+1}{2}} \frac{\Gamma(a+1)\,\zeta(a+1)}{(2\pi)^{a+1}} \right\}$$

$$= G_a(\tfrac{1}{y}) + (-1)^{\frac{a+1}{2}} \frac{\Gamma(a+1)\,\zeta(a+1)}{(2\pi)^{a+1}}$$

and hence, from (3), we obtain (4) which is our theorem. We have to modify our theorem slightly for the case a = 1. In fact when a = 1, (12) gives

$$G_1(\tfrac{1}{y}) = -\,y^2\, G_1(y) + \frac{1}{24}\, y^2 - \frac{y}{4\pi}$$

$$+ \frac{1}{2}\; \frac{2\,\Gamma(2)\,\zeta(2)}{(2\pi)^2}$$

or

(13) $$-\frac{1}{24} + G_1(\tfrac{1}{y}) = -\,y^2 \left\{ G_1(y) - \frac{1}{24} \right\} - \frac{y}{4\pi} .$$

From (13) we shall deduce that

(14) $$\frac{e^{-2\pi y} \prod_1^\infty (1 - e^{-2n\pi y})^{24}}{e^{-2\pi/y} \prod_1^\infty (1 - e^{-2n\pi/y})^{24}} = y^{-12}$$

which shows the effect of the transformation $\tau \longrightarrow -\frac{1}{\tau}$ on the denominator of $j(\tau)$. To prove (14) we write

(15) $$f(q) = q \prod_{1}^{\infty} (1-q^n)^{24} = g(\tau)$$

where

(16) $$q = e^{2\pi i \tau} \quad .$$

We have by "logarithmic differentiation",

(17) $$q \frac{f'(q)}{f(q)} = 1 - 24 \sum_{1}^{\infty} \sigma_1(n) e^{2n\pi i}$$

$$f'(q) = g'(\tau)\frac{d\tau}{dq}$$

$$= \frac{g'(\tau)}{2\pi i\ q} \quad .$$

So

(18) $$q \frac{f'(q)}{f(q)} = \frac{g'(\tau)}{2\pi i\ g(\tau)} \quad .$$

From (13),

(19) $$\tau^2 \left\{1 - 24 \sum_{1}^{\infty} \sigma_1(n) e^{2n\pi i\tau}\right\}$$
$$= \left\{1 - 24 \sum_{1}^{\infty} \sigma_1(n) e^{-\frac{2n\pi i}{\tau}}\right\} + \frac{6i\tau}{\pi}$$

or, using (17), (18), (19),

(20) $$\tau^2 \frac{g'(\tau)}{g(\tau)} \frac{1}{2\pi i} = \frac{g'(-\frac{1}{\tau})}{2\pi i\ g(\frac{1}{\tau})} + \frac{6i\tau}{\pi} \quad .$$

Hence

(21) $$\frac{g'(\tau)}{g(\tau)} = \frac{1}{\tau^2}\,\frac{g'(-\frac{1}{\tau})}{g(-\frac{1}{\tau})} - \frac{12}{\tau}$$

(22) $$\log g(\tau) = \log g(-\tfrac{1}{\tau}) - 12 \log \tau + \log k$$

where k is a constant independent of τ. Thus

(23) $$\frac{g(\tau)}{g(-\frac{1}{\tau})} = \frac{k}{\tau^{12}} .$$

Setting $\tau = iy$ this becomes

$$\frac{e^{-2\pi y}\prod_{1}^{\infty}(1 - e^{-2n\pi y})^{24}}{e^{-2\pi/y}\prod_{1}^{\infty}(1 - e^{-2n\pi/y})^{24}} = \frac{k}{y^{12}} .$$

Set $y = 1$ here to get $k = 1$; (14) is proved. From (23) [or (14)] and the case $a = 3$ of our theorem [(3) and (4)] we see at once that $j(\tau) = j(-\frac{1}{\tau})$.

§7. By combining the results of Deuring in §5 with our theorem in §1 we get the:

Theorem If p is a prime > 19 and if the class-number of $R(\sqrt{-p})$ is 1, the equation

$$X^3 - py^2 = -8$$

has a solution in rational integers and indeed with $X = \frac{1}{6}\left\{e^{\frac{\pi}{3}\sqrt{p}}\right\}$ where $\{u\}$ denotes the integer nearest to u.

Incidentally we observe that if $p > 19$ is a prime, and the class-number of $R(\sqrt{-p})$ is 1, then

$$\left\{e^{\frac{\pi}{3}\sqrt{p}}\right\}$$

is a multiple of 6!

§8. Hecke [Über die Kroneckersche Grenzformel für reele, quadratische Körper und die Klassenzahl relativ Abelscher Körper. Verhandlungen der Naturforschenden Gesellschaft in Basel, Bd. 28_2, 1917, S. 363; see also his paper "Bestimmung der Klassenzahl einer neuen Reihe von algebraischen Zahlkörpern. Gött. Nachr. 1921, S. 1.] has proved that for a real quadratic field of discriminant $\Delta > 0$, the Grenzformel takes the following form: let $\varepsilon > 1$ be a unit of the field:

(1) $$\varepsilon = \frac{u + v\sqrt{\Delta}}{2} \quad (u, v \geq 1)$$

and if α runs through a complete system of non-associated (with respect to $\pm\, \varepsilon^{\nu}$) numbers of the ideal

(2) $$\mathfrak{n} = [\alpha_1, \alpha_2], \ \alpha_1\alpha_2' - \alpha_2\alpha' = N\sqrt{\Delta} > 0, \ N = N(\mathfrak{n})$$

then we have

(3) $$N(\mathfrak{n})^s \sum_{\alpha} \frac{1}{|N(\alpha)|^s} = \frac{\lg \varepsilon^2}{\sqrt{\Delta}} \frac{1}{s-1} + \frac{2}{\sqrt{\Delta}} \phi(\mathfrak{n}) + \theta_1(s-1) + \theta_2(s-1)^2 + \ldots$$

where

(4) $$\phi(\mathfrak{n}) = E \lg \varepsilon^2 - \int_{-\lg\varepsilon}^{+\lg\varepsilon} \lg\left(\sqrt[4]{\Delta}\sqrt{\frac{\omega\cdot\bar{\omega}}{2i}}\, \eta(\omega)\,\eta(-\bar{\omega})\right) dv$$

where E is Euler's constant and

(5) $$\omega = -\frac{\alpha_2 e^v - i\alpha_2'}{\alpha_1 e^v - i\alpha_1'}, \quad \bar{\omega} = -\frac{\alpha_2 e^v + i\alpha_2'}{\alpha_1 e^v + i\alpha_1'} .$$

Herglotz (Ber. ü. a. Verhandlungen. Akad. Wiss. Leipzig. Math. Phys. Klasse, 75, 1923, 3-14) obtains some curious results from Hecke's formula. The following is a sample:

$$\varepsilon = 4 + \sqrt{15}\,, \qquad \int_0^1 \frac{\lg (1 + t^{\varepsilon})}{1 + t}\, dt = -\frac{\pi^2}{12}(\sqrt{15} - 2)$$

$$+ \lg 2 \lg (\sqrt{3} + \sqrt{5}) + \lg \frac{1 + \sqrt{5}}{2} \lg (2 + \sqrt{3})\,.$$

A direct evaluation of this definite integral is probably difficult!

VII CONSTRUCTION OF CLASS FIELDS

(Carl S. Herz, Dec. 18, 1957 revised Nov. 1965)

§0. Introduction.

The purpose of this lecture is to give some explicit constructions of class fields of imaginary quadratic fields by arithmetic means.

Let K be an algebraic number field and $\underline{H}$ the group of ideal classes of K. The absolute (Hilbert) class field of K is the class field corresponding to $\underline{H}$; it is the unique maximal unramified abelian extension H of K. In general, it is not easy to determine the structure of the group $\underline{H}$ and the explicit construction of the field H is very difficult. The problem is attacked this way. Suppose K/k is normal with Galois group Π; then H/k is normal, and by means of class field theory one can describe the Galois group Γ of H/k exactly in terms of Π, $\underline{H}$, and the action of Π on $\underline{H}$. Here we shall take $k = \underline{\underline{Q}}$ and restrict our attention to $K/\underline{\underline{Q}}$ cyclic; in this case Γ is naturally isomorphic to the semi-direct product $\Pi\underline{H}$. The rest is algebraic number theory: knowing the Galois group of $H/\underline{\underline{Q}}$ and that H/K is unramified we can go a long way towards writing H in the form $H = \underline{\underline{Q}}(\theta_1,\ldots,\theta_n)$ where the θ_i are roots of explicitly determined polynomials.

When K is a quadratic field then $\Pi \simeq \underline{\underline{Z}}_2$ and the action of Π on $\underline{H}$ is given by $\underline{k}^{\tau} = \underline{k}^{-1}$, $\underline{k} \in \underline{H}$ and τ the generator of Π. This situation is very simple indeed. Suppose L is a subfield of H which is cyclic of degree ℓ over K; then $L/\underline{\underline{Q}}$ is normal with a dihedral Galois group. It is easy to describe dihedral extensions of the rationals; thus the problem is to find such fields L which are unramified over K, and this does not require complete knowledge of the class group $\underline{H}$. The cases $\ell = 2, 3,$ or 4 are simple enough so that explicit calculations can be carried out.

In summary, if the exponent of the class group $\underline{H}$ of a quadratic number field K divides 12, the Hilbert class field H is easy to construct by arithmetic means. If K is an imaginary quadratic field we have (V, Theorem 2) $H = K(j(\underline{k}))$ where $j(\underline{k})$ is any class invariant. Except for very small discriminants the calculation of the class invariant is intractable unless $\underline{H} \simeq \mathbb{Z}_2^{t-1}$ or $\underline{H} \simeq \mathbb{Z}_4 \times \mathbb{Z}_2^{t-2}$, and in these cases the arithmetic construction of H is respectively immediate, almost immediate.

§1. Preliminaries from class field theory. Let H be the Hilbert class field of K.

LEMMA 1. *If K/k is normal then H/k is normal.*

Proof. Let σ be an isomorphism of H/k. Then H^σ is an unramified abelian extension of $K^\sigma = K$. Since H is the unique maximal abelian unramified extension of K, it follows that $H^\sigma \subset H$ and hence $H^\sigma = H$.

Let Π denote the Galois group of K/k (action on the right). Then there is a natural action of Π on $\underline{H}$, the group of ideal classes of K.

ARTIN RECIPROCITY LAW. *Let H be the class field of K corresponding to the group $\underline{H}$ of ideal classes. Then the Galois group of H/K is canonically isomorphic as a Π-module to $\underline{H}$.*

Let Γ denote the Galois group of H/k. According to the Artin Reciprocity Law, Γ is an extension of $\underline{H}$ by Π. The precise determination of Γ depends on the particular element $\bar{u} \in H^2(\Pi, \underline{H})$ describing the extension in question. According to the Weil-Shafarevitch Theorem, $\bar{u}$ is the image of u, the canonical class of K/k, under a natrual projection. In order to avoid going into details here we restrict our attention to $k = \mathbb{Q}$ (it suffices to assume that all abelian subfields of K/k are ramified over k).

COROLLARY OF SHAFAREVITCH-WEIL THEOREM. *If $K/\mathbb{Q}$ is normal with Galois* *group* Π *then the Galois group of* $H/\underline{Q}$ *is isomorphic to the semi-direct product* $\Pi\underline{H}$.

For K an imaginary quadratic field the appeal to the Shafarevitch-Weil Theorem can be avoided. We have seen in V that $H = K(j(\underline{k})) = \underline{Q}(\sqrt{d}, j(\underline{k}))$ where the class invariant $j(\underline{k})$ is a root of a polynomial with coefficients in $\underline{Q}$ which is irreducible over K. Let τ be the automorphism of $H/\underline{Q}$ given by complex conjugation, the restriction of τ to K is the non-trivial automorphism of $K/\underline{Q}$. Moreover, $(j(\underline{k}))^{\tau} = \bar{j}(\underline{k}) = j(\underline{k}^{\tau})$. Let k_1 be the unit class; then τ can also be described as the automorphism of $H/\underline{Q}$ determined by

$$\sqrt{d} \longrightarrow -\sqrt{d}, \quad j(\underline{k}_1) \longrightarrow j(\underline{k}_1) \quad .$$

The elements $\underline{k} \in \underline{H}$ are associated with automorphisms

$$\sqrt{d} \longrightarrow \sqrt{d}, \quad j(\underline{k}_1) \longrightarrow j(\underline{k}^{-1})$$

of $H/\underline{Q}$. Thus we have $2h$ distinct automorphisms $\underline{k}_1, \ldots, \underline{k}_h, \tau k_1, \ldots, \tau k_h$ of $H/\underline{Q}$ where $h = \#\underline{H}$ = class number of K. Since $[H : \underline{Q}] = 2h$, these automorphisms exhaust Γ, the Galois group of $H/\underline{Q}$. By direct calculation we have $(\sqrt{d})^{\tau^{-1}\underline{k}\tau} = \sqrt{d}$ and

$$j(\underline{k}_1)^{\tau^{-1}\underline{k}\tau} = j(\underline{k}_1)^{\underline{k}\tau} = j(\underline{k}^{-1})^{\tau} = j(\underline{k}^{-\tau})$$

which proves that $\tau^{-1}\underline{k}\tau = \underline{k}^{\tau}$. Hence Γ is isomorphic to the semi-direct product $\Pi\underline{H}$.

§2. *The genus field and the group of genera.*

Let K/k be abelian with Galois group Π, and let H be the Hilbert class field of K. The *genus field* of K/k is the maximal abelian subfield G of H/k. If Γ is the Galois group of H/k then G is the fixed field of Γ', the commutator subgroup of Γ. Since K/k is abelian, the canonical image of $\underline{H}$ in Γ must contain Γ'. Thus Γ' is canonically isomorphic as a Π-module with a subgroup $\underline{G}$ of $\underline{H}$; this subgroup is called the *principal genus*.

LEMMA 2. *The principal genus* $\underline{G}$ *is the subgroup of* $\underline{H}$ *generated by the elements* $\underline{k}^{1-\tau}$ *where* $\underline{k} \in \underline{H}$, $\tau \in \Pi$.

<u>Proof</u>. We know that Γ is canonically isomorphic to an extension of $\underline{H}$ by Π. Let $u_\tau \in \Gamma$ be a representative of $\tau \in \Pi$. The elements of Γ are identified with the products $u_\tau \underline{k}$ and $u_\tau^{-1} \underline{k}^{-1} u_\tau \underline{k} = \underline{k}^{1-\tau}$. These commutators generate Γ' since $\underline{H}$ is abelian.

The factor group $\underline{H}_\Pi = \underline{H}/\underline{G}$ is called the <u>group of genera</u> of K/k. It is canonically isomorphic under the reciprocity map with the Galois group of H/G.

The essential fact used in the construction of the genus field is given by

PROPOSITION 1. <u>Let</u> K/k <u>be abelian. The genus field of</u> K/k <u>is the unique maximal abelian extension</u> G/k <u>with the same ramification as</u> K/k.

<u>Proof</u>. Let I_k denote the group of ideals of k relatively prime to the discriminant of K/k and P_K the group of principal ideals of K, again relatively prime to the discriminant of K/k, an assumption made for all ideals in the course of this proof. An abelian extension L/k has the same ramification as K/k if $N_{K/k} P_K = N_{L/k} P_L$ as subgroups of I_k. Now let H be the Hilbert class field to K. By definition, $N_{H/K} I_H = P_K$, but by the Principal Ideal Theorem, $N_{H/K} P_H = P_K$. Since the genus field G is the maximal abelian subfield of H/k we have

$$N_{K/k} P_K = N_{K/k} N_{H/K} P_H = N_{H/k} P_H = N_{G/k} N_{H/G} P_H \subset N_{G/k} P_G$$

$$N_{K/k} P_K = N_{K/k} N_{H/K} I_H = N_{H/k} I_H = N_{G/k} N_{H/G} I_H = N_{G/k} I_G$$

and so $N_{K/k} P_K = N_{G/k} I_G = N_{G/k} P_G$. Thus G/k has the same ramification as K/k, and G/k is the class field to the subgroup $N_{K/k} P_K$ of I_k. Since for any extension L/k we have $N_{L/k} P_L \subset N_{L/k} I_L$ it follows that for any abelian extension L/k with the same ramification as K/k we have $N_{G/k} I_G \subset N_{L/k} I_L$; so L is a subfield of G.

Proposition 1 leads to a simple effective procedure for constructing genus fields. To do this we examine $N_{K/k} P_K$. Given a prime p of k let U_p

denote the group of elements of k relatively prime to p and let V_p be the subgroup of U_p consisting of those elements which are congruent to norms from K modulo arbitrarily high powers of p. Then a necessary and sufficient condition that $(\alpha) \in N_{K/k} P_K$ is

$$\alpha \in E_0 \cdot \cap V_p$$

where E_0 is the group of units of k and the intersection is taken over primes p dividing the discriminant of K/k, including possibly real infinite primes where "congruent" is to be understood as "has the same sign". In case K/k is cyclic of prime degree n it follows that

$$U_p^n \subset V_p \quad \text{and} \quad [U_p : V_p] = n$$

where U_p^n is the subgroup of U_p consisting of elements which are congruent to n-th powers modulo arbitrarily high powers of p.

The situation is particularly simple when K is a cyclic extension of the rationals of prime degree n. Ramification can only occur at primes $p \equiv 0,1 \bmod n$ since $U_p^n = U_p$ unless $n \mid \varphi(p^r)$ where $\varphi(p^r)$ is the order of $\underline{\underline{Z}}^*_{p^r}$, the group of units modulo p^r. Moreover, $\underline{\underline{Z}}^*_{p^r}$ is cyclic for $p \neq 2$. Hence

$$[U_p : U_p^n] = n \quad \text{for} \quad p \equiv 0,1 \bmod n$$

with the single exception

$$U_2/U_2^2 \simeq \underline{\underline{Z}}_2^2 .$$

At ∞ there is no ramification when n is odd, for then $U_\infty^n = U_\infty$; but $[U_\infty : U_\infty^2] = 2$.

We conclude that if $K/\underline{\underline{Q}}$ is cyclic of prime degree n and p ramifies then $V_p = U_p^n$ for all p finite or infinite except for $p = 2$, which can only occur when $n = 2$. There are three possibilities in the exceptional case: $V_2 = \{1,v\} \cdot U_2^2$ where $v = -1,5$, or 3.

It is always true that $P_{\underline{\underline{Q}}}/N_{K/\underline{\underline{Q}}} P_K \simeq \cap U_p/E_0 \cap V_p$ where p runs over the ramified primes, or, for that matter, any finite set of primes containing all the ramified primes. Using the Chinese Remainder Theorem, we obtain a direct product decomposition best written in the form (all isomorphisms being canonical).

$$P_{\underline{Q}}/N_{K/\underline{Q}}\, P_K \simeq (\cap\, U_p)/(E_0 \cap V_p) \simeq \Pi\Gamma_p$$

where $\Gamma_p = U_p/V_p$ if $E_0 \subset V_p$ and $\Gamma_p = (U_{p_0} \cap U_p)/E_0(V_{p_0} \cap V_p)$ if $E_0 \not\subset V_p$, p_0 being a fixed ramified prime such that $-1 \notin V_{p_0}$. The reason for writing the product this way is that there is a unique extension $\underline{Q}(\theta_p)/\underline{Q}$ which is cyclic of degree n and whose ramification is described by Γ_p, namely the class field of $\underline{Q}$ corresponding to Γ_p. Since the genus field G is the class field of $\underline{Q}$ corresponding to $\Pi\Gamma_p$, it must be the compositum of the $\underline{Q}(\theta_p)$. What we have proved is

THEOREM 1. Let $K/\underline{Q}$ be cyclic of prime degree n. The genus field of $K/\underline{Q}$ is $G = K(\theta_1,\dots,\theta_t)$ where $p_1,\dots,p_t$ are the distinct prime divisors of the discriminant d (excluding $p_0 = \infty$ if K is imaginary quadratic and an arbitrary $p_0 \equiv -1 \bmod 4$ if K is real quadratic) where for each p_i, $\underline{Q}(\theta_i)$ is cyclic of degree n over $\underline{Q}$ and specified this way:

(i) for n odd and p given (so $p \equiv 0,1 \bmod n$), $\underline{Q}(\theta)$ is the unique cyclic extension of degree n which ramifies only at p,

(ii) for $n = 2$ and $p \equiv 1 \bmod 4$, $\theta = \sqrt{p}$,

(iii) for $n = 2$ and $p \equiv -1 \bmod 4$, $\theta = \sqrt{-p}$ or $\theta = \sqrt{p_0 p}$ according to the imaginary or real case,

(iv) for $n = 2$ and $p = 2$, so $K = \underline{Q}(\sqrt{d})$, the cases are

$\theta = \sqrt{2}$ if $d \equiv 8 \bmod 32$

$\theta = \sqrt{-1}$ or $\sqrt{p_0}$ if $d \equiv 4 \bmod 8$ according to $d < 0$ or $d > 0$

$\theta = \sqrt{-2}$ or $\sqrt{2p_0}$ if $d \equiv -8 \bmod 32$ according to $d < 0$ or $d > 0$.

It may not be obvious how we arrived at (iv). To see it one observes that if 2 ramifies then $K = \underline{Q}(\sqrt{d_0})$ where the discriminant is $4d_0$. Then

$1 - d_0 \in V_2$ since $1 - d_0$ is the norm of $1 + \sqrt{d_0}$. If $d_0 \equiv 2 \bmod 8$ then $-1 \in V_2$ and $\Gamma_2 = U_2/V_2$ corresponds to $\underline{Q}(\sqrt{2})$. The remaining possibilities are $d_0 \equiv -2 \bmod 8$ and $d_0 \equiv -1 \bmod 4$. In the former case $3 \equiv 1 - d_0 \bmod 8$ so $3 \in V_2$; in the latter $5 \equiv 1 - 4d_0 \bmod 8$ so $5 \in V_2$. In neither of these cases do we have $-1 \in V_2$; hence $\Gamma_2 = U_{p_0} \cap U_2/E_0(V_{p_0} \cap V_2)$. If $d < 0$ we can take $p_0 = \infty$ and the class field corresponding to Γ_2 in $\underline{Q}(\sqrt{-1})$ or $\underline{Q}(\sqrt{-2})$ according to whether $5 \in V_2$ or $3 \in V_2$. When $d > 0$ we must use $\underline{Q}(\sqrt{p_0})$ and $\underline{Q}(\sqrt{2p_0})$ instead where $p_0 \equiv -1 \bmod 4$.

REMARK. In the classical treatment of real quadratic fields K one considers H^+, the maximal abelian extension of K which is unramified except at ∞. One puts G^+ the maximal abelian subfield of $H^+/\underline{Q}$ and then $G^+ = \underline{Q}(\sqrt{p_1},\ldots,\sqrt{p_s})$ where $p_1,\ldots,p_s$ are the distinct prime divisors of the discriminant. Therefore either $G^+ = G$ or $G^+ = G(\sqrt{p_0})$ according to whether or not there is a prime $p_0 \equiv -1 \bmod 4$ dividing the discriminant.

The Galois group of $G/\underline{Q}$ is Γ/Γ' and, as we have just seen, this is isomorphic to $\prod_{p|d} \Gamma_p$. The reciprocity law makes this isomorphism explicit. Assume that the action of Π has been extended to G. Then Π acts on each $\underline{Q}(\theta_p)$; we shall write $p = p'$ if Π acts non-trivially, $p = p''$ if Π acts trivially. Let q be a prime such that $q \nmid d$, $q \notin V_{p'}$, $q \in V_{p''}$ (assume $q > 0$ and $q \in V_{p_0}$ in the quadratic cases). Then q determines a generator $\gamma_{p'}$ of $\Gamma_{p'}$. The reciprocity law for the extension $\underline{Q}(\theta_{p'})/\underline{Q}$ says $\gamma_{p'}\left(\frac{Q(\theta_{p'})/Q,q}{q}\right) = 1$. The interpretation of the reciprocity symbol gives the fact that q remains prime in $\underline{Q}(\theta_{p'})/\underline{Q}$ and $\gamma_{p'}$ induces the Frobenius automorphisms on the residue class extension at q. This determines $\gamma_{p'}$ uniquely. It follows that $\left(\frac{G/\underline{Q},q}{q}\right) = (\Pi\gamma_{p'})^{-1}$. On the other hand q must remain prime in $K/\underline{Q}$ since its divisors in $G/\underline{Q}$ are fixed under Π. Hence $\left(\frac{K/\underline{Q},q}{q}\right)^{-1}$ is the generator of

Π which induces the Frobenius automorphism on the residue class extension at q. Clearly $(\frac{K/\underline{Q},q}{q})$ is the image of $(\frac{G/\underline{Q},q}{q})$ under the natural projection of the Galois group of $G/\underline{Q}$ onto the Galois group of $K/\underline{Q}$. Hence $\tau = \prod_{p'|d} \gamma'_p$ is a generator of Π. If the action of Π on G has not been given in advance it suffices to choose q such that q remains prime in $K/\underline{Q}$ and determine the p' as those $p|d$ such that $q \notin V_p$.

To summarize, we state

THEOREM 2. _Let_ $K/\underline{Q}$ _be cyclic of prime degree_ n _with Galois group_ Π. _Given an action of_ Π _on the genus field_ $G = \underline{Q}(\theta_1,\ldots,\theta_t)$ _and the reciprocity law identification of_ Γ_{p_i} _with the Galois group of_ $\underline{Q}(\theta_i)/\underline{Q}$, _there is a choice of generators_ γ_p _for_ Γ_p _such that_ $\tau = \prod_{p'} \gamma_{p'}$ _generates_ Π _and the group of genera is_ $\underline{H}_\Pi = (\prod_{p|d} \Gamma_p)/\Pi$, _where_ p' _denotes those_ p_i _such that_ Π _acts non-trivially on_ θ_i.

For K imaginary quadratic one takes τ to be complex conjugation on H, hence also on G. Then the p' are the $p|d$ such that $p \equiv -1 \bmod 4$ and 2 if $d \equiv 4 \bmod 8$ or $d \equiv -8 \bmod 32$.

The concrete isomorphism of $\underline{H}_\Pi$ with $(\prod_{p|d} \Gamma_p)/\Pi$ is made this way. Each genus can be represented by a prime $\underline{q}$ of K, but since conjugates belong to the same genus it suffices to specify the rational prime q which $\underline{q}$ divides. The groups in question are isomorphic to $\underline{Z}_n^{t-1}$; so it remains only to give rational primes $q_1,\ldots,q_{t-1}$ representing generators of $\underline{H}_\Pi$ in correspondence with certain elements of $\prod_{p|d} \Gamma_p$. Given γ_p as an automorphism of $G/\underline{Q}$, let $\bar{\gamma}_p$ be the automorphism of $K/\underline{Q}$ obtained from the natural projection of the Galois group of $G/\underline{Q}$ onto the Galois group of $K/\underline{Q}$. Take p_t so that $\bar{\gamma}_{p_t}$ is non-trivial (this is no restriction at all). Since Π is cyclic of prime degree we can find integers $e_1,\ldots,e_{t-1}$ such that $\bar{\gamma}_{p_i} \bar{\gamma}_{p_t}^{e_i} = 1$. Now consider the γ_p as elements of the groups Γ_p defined by certain congruences in $\underline{Q}$ and choose primes $q_1,\ldots,q_{t-1}$ ($q_i > 0$ and $q_i \in V_{p_0}$ in the quadratic cases) such that

$$q_i \in \gamma_{p_i} V_{p_i} \cap \gamma_{p_t}^{e_{p_i}} V_{p_t} \cap \bigcap_{j\neq i,t} V_{p_j} .$$

THEOREM 3. *The canonical isomorphism* $(\prod_{p|d} \Gamma_p)/\Pi \longrightarrow H_\Pi$ *is completely specified by* $\gamma_{p_i} \gamma_{p_t}^{e_i} \longrightarrow q_i$, $i = 1,\ldots,t-1$. *Here* $(\prod_{p|d} \Gamma_p)/\Pi$ *is to be viewed as the Galois group of* G/K *and* H_Π *as the group of genera of* K.

Proof. The $\gamma_{p_i} \gamma_{p_t}^{e_i}$ generate a subgroup Δ of $\prod_{p|d} \Gamma_p$. For $\delta \in \Delta$ we have $\bar{\delta} = 1$, but $\bar{\tau} \neq 1$ where τ generates Π. Thus $\Delta \cap \Pi = \{1\}$, and $\Delta \simeq Z_n^{t-1} \simeq (\prod_{p|d} \Gamma_p)/\Pi$. Therefore we may identify Δ with the Galois group of $G/K = K(\theta_1,\ldots,\theta_{t-1})/K$. According to the reciprocity law, q_i splits in K/Q and in each $Q(\theta_j)/Q$ for $j \neq i,t$ but does not split in $Q(\theta_i)/Q$. It follows that $\mathfrak{q}_i$ splits completely in $K(\theta_1,\ldots,\theta_{i-1},\theta_{i+1},\ldots,\theta_{t-1})/K$ and remains prime in $K(\theta_i)/K$, where $\mathfrak{q}_i$ is any prime of K dividing q_i. The reciprocity law demands that we associate $\mathfrak{q}_i$ with the automorphism of $K(\theta_1,\ldots,\theta_{t-1})/K$ which leaves θ_j fixed for $j \neq i$ and which induces the Frobenius automorphism in the residue-class extension at $\mathfrak{q}_i$ of $K(\theta_i)/K$. Since q_i splits in K/Q, the residue-class extension at $\mathfrak{q}_i$ of $K(\theta_i)/K$ is the same as the residue-class extension at q_i of $Q(\theta_i)/Q$. The automorphism γ_i of $Q(\theta_i)/Q$ corresponds to $\gamma_i \gamma_t^{e_i} \in \Delta$ when Δ is taken as the Galois group of $K(\theta_1,\ldots,\theta_{t-1})/K$.

The above three theorems give a complete description of the theory of genera for K/Q cyclic of prime degree n. This is equivalent to the theory of the representation of integers by certain homogeneous forms of degree n in n rational variables with integral coefficients. The explicit computations are tedious when $n > 2$. In the quadratic case the computations are nearly trivial, especially since there is no choice of generators to worry about.

For contrast, we shall give on illustration of Theorems 1, 2, and 3 in the cubic case.

<u>Illustration</u>. Take $K = \underline{Q}(\theta)$ where $\theta^3 - 21\theta - 28 = 0$. Ramification occurs at $p_1 = 3$, $p_2 = 7$. The genus field is, according to Theorem 1, $G = \underline{Q}(\theta_1, \theta_2)$ where $\theta_1^3 - 3\theta_1 - 1 = 0$ and $\theta_2^3 - 3\cdot 7\theta_2 - 7 = 0$. We see that $q = 5$ remains prime in $\underline{Q}(\theta)$, $\underline{Q}(\theta_1)$, and $\underline{Q}(\theta_2)$. Hence, by Theorem 2, $\tau = \gamma_3\gamma_7$ where γ_3 corresponds to elements $\equiv \pm 4 \bmod 9$ and γ_7 to elements $\equiv \pm 2 \bmod 7$. It necessarily follows that $\bar{\gamma}_3 = \tau^2$, $\bar{\gamma}_7 = \tau^2$. According to Theorem 3 the non-principal genera are given by $59 \longleftrightarrow \gamma_3\gamma_7^2$ and $2 \longleftrightarrow \gamma_3^2\gamma_7$. We have $\underline{H}_\Pi \simeq \underline{\underline{Z}}_3$. (Also $\underline{H} = \underline{H}_\Pi$.) The theory states that $m \in U_3 \cap U_7$ is the norm of a principal ideal, i.e. there exist $x,y,z \in \underline{Q}$ such that

$$m = x^3 + 7\{4y^3 + 16\cdot 7z^3 + 6x^2z - 3xy^2 + 9\cdot 7xz^2 - 12\cdot 7yz^2\},$$

iff $m \equiv \pm 1 \bmod 9$ and $m \equiv \pm 1 \bmod 7$. The cubic reciprocity law states that for a prime $q \neq 3,7$ the congruence

$$28 \equiv w^3 - 21w \bmod q$$

has a solution iff q corresponds to $\gamma_3^a\,\gamma_7^b$ where $\bar{\gamma}_3^a\,\bar{\gamma}_7^b = 1$, i.e. $q \equiv \pm 2^k \bmod 63$. This is equivalent to q being the norm of an ideal. Thus 2 and 59 are norms of non-principal ideals while $2\cdot 59$ is the norm of a principal ideal. The presentation of τ, γ_3, and γ_7 as automorphisms of the appropriate fields is too tedious.

Let us return to the quadratic case. Let d be the discriminant of a quadratic field $K = \underline{Q}(\sqrt{d})$ and m a positive integer relatively prime to d (also assume $(\frac{m}{p_0}) = 1$). We have seen that m is the norm of a principal ideal in K, i.e.

$$x^2 - dy^2 = 4m$$

has solutions $x,y \in \underline{Q}$ iff $(\frac{m}{p}) = 1$ for each odd $p|d$ and $m \equiv 1,v \bmod 8$ if $2|d$. There is one class in each genus, i.e. $\underline{H} = \underline{H}_\Pi$, iff whenever solutions $x,y \in \underline{Q}$ exist then solutions $x,y \in \underline{\underline{Z}}$ exist. Here $(\frac{m}{p}) = +1$ if m is a quadratic residue mod p, and if m is a prime, the splitting of m in $K/\underline{Q}$

is determined by the classical quadratic reciprocity law.

Example 1. Take $d = -84$. Then $H = G = \underline{\underline{Q}}(\sqrt{-1}, \sqrt{-3}, \sqrt{-7})$. We have $\tau = \gamma_2\gamma_3\gamma_7$. Generators for $\underline{H}_\Pi$ are given by

$$19 \longleftrightarrow \gamma_2\gamma_7 \qquad 5 \longleftrightarrow \gamma_3\gamma_7 \ .$$

The equation $x^2 + 21y^2 = m$ has integral solutions whenever it has rational solutions.

Example 2. Take $d = -87$. Then $G = \underline{\underline{Q}}(\sqrt{-3}, \sqrt{29})$. We have $\tau = \gamma_3$ and a generator for $\underline{H}_\Pi$ is given by $2 \longleftrightarrow \gamma_3\gamma_{29}$. Notice that 7 corresponds to the principal genus, and, indeed, the equation $x^2 + 87y^2 = 4\cdot 7$ has the solution $x = 5/2$, $y = 1/2$ but it does not have integral solutions. Hence the primes dividing 7 give non-principal classes in the principal genus. On the other hand $16^2 + 87\cdot 1^2 = 7^3$. Therefore if $\underline{P}_7$ is a prime of K dividing 7 then $\underline{P}_7$ is in the principal genus and $\underline{P}_7^3$ is in the principal class. In fact $\underline{H} \simeq \underline{\underline{Z}}_2 \times \underline{\underline{Z}}_3$. We shall see later that $H = Q(\sqrt{-3}, \sqrt{29}, \varepsilon^{1/3})$ where $\varepsilon^2 - 5\varepsilon - 1 = 0$.

Example 3. Take $d = -39$. Then $G(\sqrt{-3}, \sqrt{13})$. We have $\tau = \gamma_3$ and a generator for $\underline{H}_\Pi$ is given by $2 \longleftrightarrow \gamma_3\gamma_{13}$. Let $\underline{P}_2$ be a prime of K dividing 2. Obviously $\underline{P}_2^2$ is in the principal genus, but $\underline{P}_2^2$ is not principal. To see this, observe that $\underline{P}_2 \neq \underline{P}_2^\tau$ since 2 is unramified. Hence $\underline{P}_2^2 \neq (2) = \underline{P}_2^{1+\tau}$; so if $\underline{P}_2^2$ were principal there would be a solution to $x^2 + 39y^2 = 4\cdot 2^2$ with $y \neq 0$, $y \in \underline{\underline{Z}}$. Here $\underline{H} \simeq \underline{\underline{Z}}_4$.

Example 4. Take $d = 40$. Here $G = \underline{\underline{Q}}(\sqrt{2}, \sqrt{5})$. We have $\tau = \gamma_5$. The generator for $\underline{H}_\Pi$ is given by $3 \longleftrightarrow \gamma_2\gamma_5$. The equation $x^2 - 10y^2 = m$ has integral solutions whenever it has rational solutions and there is no need to take $m > 0$ since $3^2 - 10\cdot 1^2 = -1$.

Example 5. Take $d = 21$. Here $G = K$. The equation $x^2 - 21y^2 = 4m$ has integral solutions for all m such that $(\frac{m}{3}) = 1$ and $(\frac{m}{7}) = 1$. Therefore $x^2 - 21y^2 = \pm 4m$ has solutions whenever $(\frac{m}{3})(\frac{m}{7}) = 1$. If we impose the restriction

$m > 0$ there is a change: the ideal (17) is the norm of a principal ideal but 17 is not a norm. In this example $H = K$, but $H^+ = \underset{=}{Q}(\sqrt{-3}, \sqrt{-7})$.

Example 6. Take $d = 136$. Here $G = \underset{=}{Q}(\sqrt{2}, \sqrt{17})$. There is a choice of $\tau = \gamma_2$ or $\tau = \gamma_{17}$. The equation $x^2 - 34y^2 = m$ has rational solutions for all $m \equiv \pm 1 \bmod 8$, $\equiv \pm 2^k \bmod 17$. Whenever $x^2 - 34y^2 = m$ has rational solutions then $x^2 - 34y^2 = \pm m$ has integral solutions but not necessarily with the $+$ sign. The simplest example is $(5/3)^2 - 34(1/3)^2 = -1$; there is no integral solution to $x^2 - 34y^2 = -1$. In this example $\underline{H}^+ \simeq \underset{=}{Z}_4$ while $\underline{H}^+_{\Pi} = \underline{H}_{\Pi} = \underline{H} \simeq \underset{=}{Z}_2$.

§3. Structure of the group of ideal classes.

Let $K/\underset{=}{Q}$ be an abelian extension of order n. Recall that $\underline{H}$ denotes the group of ideal classes of K and $\underline{G}$ the principal genus, i.e. the subgroup of $\underline{H}$ generated by the $\underline{k}^{1-\tau}$, $\tau \in \Pi$ where Π is the Galois group of $K/\underset{=}{Q}$. Put $N = \Sigma_{\tau\in\Pi}\, \tau$. Then for any $\underline{k} \in \underline{H}$ we have that $\underline{k}^{n-N} \in \underline{G}$. On the other hand, $\underline{k}^N$ corresponds to the ideal class of the norm of an ideal in $\underline{k}$ and this, being an ideal of $\underset{=}{Q}$, is principal. It follows that $\underline{H}^n$, the group of n-th powers in $\underline{H}$ is a subgroup of $\underline{G}$. Hence the group of genera $\underline{H}_{\Pi} = \underline{H}/\underline{G}$ is a subgroup of $\underline{H}/\underline{H}^n$.

In case n is a prime we know $\underline{H}_{\Pi}$ as the Galois group of G/K where the genus field G is given by Theorem 1. The fundamental numerical invariant here was t where

t = number of distinct prime factors of the discriminant d of $K/\underset{=}{Q}$ (excluding one prime factor $p_0 \equiv -1 \bmod 4$ if K is real quadratic).

THEOREM 4. Let $K/\underset{=}{Q}$ be cyclic of prime degree n. Then $\underline{H}_{\Pi} \simeq \underset{=}{Z}_n^{t-1}$. For $n = 2$, $\underline{H}_{\Pi} \simeq \underline{H}/\underline{H}^2$, for $n > 2$, $\underline{H}_{\Pi}$ is a quotient group of $\underline{H}/\underline{H}^n$ which is non-trivial if $\underline{H}^n \neq \underline{H}$.

COROLLARY. Let h be the class number of K. Then $n|h$ iff $t > 1$.

Proof of Theorem 4. According to Theorem 1 the Galois group of $G/\underset{=}{Q}$

is the direct product of t cyclic groups of order n. This shows that $\underline{H}_{\Pi} \simeq \underline{\underline{Z}}_n^{t-1}$. When $n = 2$, $N = 1 + \tau$. Hence $k^{1-\tau} = k^2$, since $k^{\tau} = k^{-1}$. Therefore $\underline{G} = \underline{H}^2$ for $n = 2$. When $n > 2$ we cannot conclude that $\underline{G} = \underline{H}^n$ but Π acts on $\underline{H}$ leaving $\underline{H}^n$ invariant. Thus the cyclic group of order n, Π, acts on the n-group $\underline{H}/\underline{H}^n$. If $\underline{H} \neq \underline{H}^n$ there must be a non-trivial fixed point of Π in $\underline{H}/\underline{H}^n$; this implies that $\underline{G} \neq \underline{H}$.

To elaborate the last statement further, let $\underline{H}^{\Pi}$ denote the subgroup of $\underline{H}$ composed of ambiguous ideal classes, ideal classes $\underline{k}$ which are invariant under Π, i.e. $\underline{k}^{\tau} = \underline{k}$. The sequence

$$1 \longrightarrow \underline{H}^{\Pi} \longrightarrow \underline{H} \xrightarrow{1-\tau} \underline{H} \longrightarrow \underline{H}_{\Pi} \longrightarrow 1$$

is exact; hence $\underline{H}^{\Pi}$ and $\underline{H}_{\Pi}$ have the same order (observe that $\underline{G} \simeq \underline{H}/\underline{H}^{\Pi}$). For $\underline{k} \in \underline{H}^{\Pi}$, however, $\underline{k}^n = \underline{k}^N = 1$. Since n is a prime, it follows that $\underline{H}^{\Pi} \simeq \underline{H}_{\Pi}$. One may argue directly that if $\underline{H}$ contains an element of order n then $\underline{H}^{\Pi}$, the group of fixed points, must be non-trivial.

REMARK. The old-fashioned treatment of genera of quadratic fields was based on the isomorphism of $\underline{H}^{\Pi}$ with $\underline{H}_{\Pi}$. One proved directly that $\underline{H}^{\Pi} \simeq \underline{\underline{Z}}_n^{t-1}$. For imaginary quadratic fields the analysis of $\underline{H}^{\Pi}$ is quite elementary and yields the construction of the genus field G in a very straightforward fashion. It gives no information, however, about the representation of $\underline{H}_{\Pi}$ as automorphisms of G/K. In the real quadratic case, the analysis of $\underline{H}^{\Pi}$ involves consideration of the units of K, but this is ultimately irrelevant.

For $n > 2$ the action of Π on $\underline{H}$ may be complicated. For $n = 2$ it is just $\underline{k}^{\tau} = \underline{k}^{-1}$. Therefore we have immediately

THEOREM 5. *Let K/$\underline{Q}$ be a quadratic field and L an unramified abelian extension of K. Then L is normal over $\underline{Q}$. Moreover if L/K is cyclic then L/$\underline{Q}$ is dihedral.*

A consequence of Theorem 5 is that when K is quadratic, H is the compositum of dihedral extensions of $\underline{\underline{Q}}$. These are, in principal, manageable as the splitting fields of polynomials with rational coefficients.

Theorem 4 gives the number of 2-primary components of the group of ideal classes of a quadratic field. For ℓ an odd prime, the question of whether ℓ / h where h is the class number is best handled by examining the possibility of the existence of a cyclic unramified extension L/K of degree ℓ. I shall treat this topic in §4 for the case $\ell = 3$, cf. Theorem 6.

In case $\underline{H} \simeq \underline{\underline{Z}}_2^{t-1}$ then $\underline{H} = \underline{H}_\Pi$ and so $H = G$. The absolute class field H is then given by Theorem 1; all that is involved is the simultaneous adjunction of square roots of rational integers. The table below shows that for $|d|$ small these trivial cases are quite common. (The first exception for $d > 0$ is $d = 145$, $\underline{H} \simeq \underline{\underline{Z}}_4$.) One knows that there are only finitely many quadratic fields with a given $\underline{H}$ as class group. I do not know which finite commutative groups are realizable as the group of ideal classes of an imaginary quadratic field.

Let $K = \underline{Q}(\sqrt{d})$ be a quadratic field of discriminant $d < 0$. Below we list the structures of the groups of ideal classes for all $d \geq -168$ and also $d = 4d_0$ with $d_0 > -100$ when these are outside the list of $d \geq -168$ but give a group other than $\underline{Z}_2^{t-1}$.

$\underline{H}$	$-d$
(1)	3, 4, 7, 8, 11, 19, 43, 67, 163, ?
$\underline{Z}_2$	15, 20, 24, 35, 40, 51, 52, 88, 91, 123, 148, ...
$\underline{Z}_2^2$	84, 120, 132, 168, ...
$\underline{Z}_2^3$	420, ...
$\underline{Z}_4$	39, 55, 56, 68, 111, 136, 155, ...
$\underline{Z}_8$	95, 164, ...
$\underline{Z}_2 \times \underline{Z}_4$	260, ...
$\underline{Z}_3$	23, 31, 59, 107, 139, ...
$\underline{Z}_2 \times \underline{Z}_3$	87, 104, 116, 152, ...
$\underline{Z}_4 \times \underline{Z}_3$	356, ...
$\underline{Z}_5$	47, 79, 103, 127, 131, ...
$\underline{Z}_2 \times \underline{Z}_5$	119, 143, 159, ...
$\underline{Z}_7$	71, ...
$\underline{Z}_2 \times \underline{Z}_7$	151, ...
$\underline{Z}_{11}$	167, ...

§4. Unramified cubic extensions.

Suppose L/K is cyclic of degree 3. Choose a generator σ for the Galois group of L/K and let ω be a primitive cube root of unity. We suppose $\omega \notin K$. Then $L(\omega)/K$ is cyclic of degree 6. We extend σ to an automorphism of $L(\omega)/K$ by putting $\omega^{\sigma} = \omega$. Let ρ denote the unique automorphism of $L(\omega)/K$ which has order 2. There is an element $\xi \in L(\omega)$ such that $\xi^{\sigma-1} = \omega$; ξ is uniquely determined by the choice of the pair (σ,ω) up to multiplication by $\beta \in K($

Define quantities a, e, and θ by

$$\theta = \xi + \xi^{\rho}, \qquad a = \xi^{1+\rho}, \qquad e = \xi^{2-\rho} + \xi^{2\rho-1} .$$

It is easy to check that $a,e \in K$, $\theta \in L$. Indeed

$$L = K(\theta) \text{ where } \theta^3 - 3a\theta - ae = 0 .$$

Now suppose $K/\underline{\underline{Q}}$ is quadratic and $L/\underline{\underline{Q}}$ is dihedral. We can extend the non-trivial automorphism of $K/\underline{\underline{Q}}$ to an automorphism τ of $L(\omega)/\underline{\underline{Q}}$ with the properties

$$\omega^{\tau} = \omega^2 , \quad \tau^2 = 1 , \quad \rho\tau = \tau\rho , \quad \sigma\tau = \tau\sigma^2 .$$

Put $\beta = \xi^{\tau-1}$. Then $\beta^{\sigma} = \xi^{\tau\sigma-\sigma} = \xi^{\sigma^2\tau}\,\xi^{-\sigma} = (\omega^2\xi)^{\tau}(\omega^{-1}\xi^{-1}) = (\omega\xi^{\tau})(\omega^{-1}\,\xi^{-1}) = \xi^{\tau-1}$. Hence $\beta \in K(\omega)$, and we are free to replace ξ by $(1+\beta)\xi$ if $\beta \neq -1$; we replace ξ by $(\omega-\omega^2)\xi$ if $\beta = -1$. With these replacements we may assume that $\xi = \xi^{\tau}$ and ξ is determined uniquely up to a factor $\gamma \in K_1$ where $K = \underline{\underline{Q}}(\sqrt{d})$, $K_1 = \underline{\underline{Q}}(\sqrt{-3d})$. Moreover $a,e \in \underline{\underline{Q}}$.

If L/K is unramified then $L(\omega)/K(\omega)$ is unramified. Since $\xi^3 \in K(\omega)$, it follows that for a prime $\underline{P}$ of $K(\omega)$ if $\underline{P}|\xi^3$ then $\underline{P}^3|\xi^3$. On the other hand $\xi^3 \in K_1$; hence if $\underline{p}$ is a prime of K_1 and $\underline{p}|\xi^3$ then $\underline{p}^3|\xi^3$. There is no loss in generality in assuming that ξ^3 is an integer of K_1 which is not divisible by the cube of any integer (other than a unit) in K_1. The norm from K_1 to $\underline{\underline{Q}}$ of ξ^3 is $\xi^{3(1+\rho)} = a^3$. The conclusion is that the rational prime p divides $a \in \underline{\underline{Z}}$ only if p factors in K_1 into non-principal prime ideals whose cubes are principal.

The above reasoning leads to

THEOREM 6. *Let* $K = \underline{Q}(\sqrt{d})$ *be an imaginary quadratic field with class number* h. *Put* $K_1 = \underline{Q}(\sqrt{-3d})$. *Let* h_1 *be the class number of* K_1, ε *the fundamental unit of* K_1, a *the norm of* ε, *and* e *the trace of* ε. *Then* $3|h$ *iff at least one of the following conditions holds*

(i_1) $a = +1, \quad e \equiv \pm 2 \bmod 27$

(i_2) $e \equiv 0 \bmod 9$

(i_3) $a = -1, \quad e \equiv \pm 4 \bmod 9$

(ii) $3|h_1$.

If (i) *holds then* $K(\theta)/K$ *is an unramified cyclic cubic extension of* K *where* $\theta^3 - 3a\theta - ae = 0$.

REMARK. Cases (i_2) and (i_3) can only occur if $3|d$.

Proof. If $3|h$ then K has an unramified cyclic cubic extension $L = K(\theta)$ where $\theta = \xi + \xi^\rho$ and $\xi^3 \in K_1$. As we saw above, either (ii) holds or ξ^3 is a unit. If ξ^3 is a unit then, modifying ξ by a factor $\pm\varepsilon^k$ we can suppose that $\xi^3 = \varepsilon$ or ε^ρ. The extension $K(\theta)/K$ where $\theta^3 - 3a\theta - ae = 0$ can only ramify at primes dividing 3 in K. If $3 \nmid d$ then 3 does not ramify in $K/\underline{Q}$, and hence $K(\theta)/K$ is ramified iff $\underline{Q}(\theta)/\underline{Q}$ ramifies at 3. In turn $\underline{Q}(\theta)/\underline{Q}$ is unramified at 3 iff $e \equiv \pm 2 \bmod 27$ (because when $3 \nmid d$ we must have $a = 1$ and $3 \nmid e$). When $3|d$ the analysis is more complicated, but in all cases condition (i) is necessary and sufficient that $K(\theta)/K$ be unramified. Finally, if $3|h_1$ then K_1 has an unramified cyclic cubic extension, but a real quadratic field $\underline{Q}(\sqrt{d_1})$ can have such an extension only if the class number of the imaginary quadratic field $\underline{Q}(\sqrt{-3d_1})$ is divisible by 3 since $\underline{Q}(\sqrt{-3d_1})$ has no units other than ± 1.

Examples.

(i_1) This is the case most frequently encountered, e.g. $d = -23$, $e = 25$;

$d = -31$, $e = 29$; $d = -116$, $e = 56$; $d = -152$, $e = 2050$; $d = -1979$, $e = 79$. The case $d = -107$ is more interesting. One has $N(\frac{11+9\sqrt{-107}}{2}) = 13^3$ and $h = 3$. Also $h_1 = 3$, $\varepsilon = 215 + 12\sqrt{321}$. Using (i_1) we get $H = \underline{Q}(\sqrt{-107},\theta)$ where $\theta^3 - 3\theta - 430 = 0$. Note that $N(\frac{17+\sqrt{321}}{2}) = (-2)^3$ but $\xi^3 = (17 + \sqrt{321})/2$ does not yield an unramified extension since $\xi^6 \not\equiv 1 \bmod \underline{P}_3^3$ where $\underline{P}_3$ is the prime dividing 3 in K_1. The absolute class field of K_1 is $H_1 = \underline{Q}(\sqrt{321},\theta_1)$ where $\theta_1^3 - 3\cdot 13\theta_1 - 11 = 0$.

(i_2) $d = -3\cdot 7\cdot 11$. Here $\underline{H} \simeq \underline{\underline{Z}}_2^2 \times \underline{\underline{Z}}_3$. We have $h_1 = 1$ and $\varepsilon = (9 + \sqrt{77})/2$. The absolute class field in $H = \underline{Q}(\sqrt{-3}, \sqrt{-7}, \sqrt{-11}, \varepsilon^{1/3})$.

(i_3) $d = -3\cdot 29$. Here $\underline{H} \simeq \underline{\underline{Z}}_2 \times \underline{\underline{Z}}_3$, $h_1 = 1$, $\varepsilon = (5 + \sqrt{29})/2$. The absolute class field is $H = \underline{Q}(\sqrt{-3}, \sqrt{29}, \varepsilon^{1/3})$.

(ii) $d = -3\cdot 229$. One has $N(26 + 3\sqrt{-687}) = 19^3$, hence $3|h$. Also $2|h$ and, in fact, $\underline{H} \simeq \underline{\underline{Z}}_2 \times \underline{\underline{Z}}_3$. We have $K_1 = \underline{Q}(\sqrt{229})$, $\varepsilon = (15 + \sqrt{229})/2$, so none of the cases (i) apply. On the other hand $N(6 + \varepsilon) = 5^3$ and $(6 + \varepsilon)^2 \equiv 1 \bmod 9$. The absolute class field to K is $H = (\sqrt{-3}, \sqrt{229}, (6 + \varepsilon)^{1/3})$. At the same time, since $h_1 = 3$, the absolute class field to K_1 is $H_1 = \underline{Q}(\sqrt{229}, \theta_1)$ where $\theta_1^3 - 3\cdot 19\theta_1 - 52 = 0$.

§5. Unramified cyclic extensions of degree 4.

Let $K = \underline{Q}(\sqrt{d})$ be a quadratic field of discriminant d. Suppose L/K is an unramified cyclic extension of degree 4. We know that the Galois group of $L/\underline{Q}$ is given by generators σ and τ, where K is the fixed field of σ, satisfying the relations $\sigma^4 = 1 = \tau^2$, $\sigma\tau = \tau\sigma^3$. Let F be the fixed field of σ^2. Then F is an unramified quadratic extension of K; hence $F \subset G$ where G is the genus field. It follows from Theorem 1 that $F = \underline{Q}(\sqrt{d_+}, \sqrt{d_-})$ where $d = d_+d_-$; the factors are distinguished by

$$\sqrt{d_+}^{\,\tau} = +\sqrt{d_+}\ , \quad \sqrt{d_-}^{\,\tau} = -\sqrt{d_-}\ .$$

(If $d < 0$ this means $d_+ > 0$ and $d_- < 0$.) Take R the fixed field of τ.

Then R is quadratic over $\underline{Q}(\sqrt{d_+})$; so we can write $R = \underline{Q}(\sqrt{d_+}, \sqrt{\mu})$ where $\mu \in \underline{Q}(\sqrt{d_+})$ and $\sqrt{\mu}^{\sigma^2} = -\sqrt{\mu}$. On the other hand $R/\underline{Q}$ is not normal; hence $\mu \notin \underline{Q}$ and $L = \underline{Q}(\sqrt{d_-}, \sqrt{\mu})$. One finds easily that $\sqrt{\mu}^{1+\sigma}$ reverses sign under the actions of both σ and τ. It follows that $\sqrt{\mu}^{1+\sigma} = m\sqrt{d_-}$ where $m \in \underline{Q}$. The result is

$$N\mu = \mu^{1+\sigma} = m^2 d_-$$

where N denotes the norm from $\underline{Q}(\sqrt{d_+})$ to $\underline{Q}$.'

There is obviously no loss in generality in assuming that μ is a square-free integer of $\underline{Q}(\sqrt{d_+})$. This fixes μ up to multiplication by the square of a unit if $\underline{Q}(\sqrt{d_+})$ has all its ideals principal. In general, what one can say is given in

LEMMA 3. *Let p be a rational prime; if $p|d_-$ then p splits in $\underline{Q}(\sqrt{d_+})$. Write $d_- = d_0$ or $d_- = 4d_0$, where $4 \nmid d_0$. The factorization of μ in $\underline{Q}(\sqrt{d_+})$ is $(\mu) = \underline{m}^2\underline{d}$ where $N\underline{d} = (d_0)$ and $\underline{m}$ has no principal ideal factors.*

Proof. Let $\underline{p}$ be a prime of $\underline{Q}(\sqrt{d_+})$ and suppose $\underline{p}^k$ exactly divides μ. In order that $F(\sqrt{\mu})$ be unramified over F it is necessary that $\underline{p}^k$ be a square in F. Hence k must be even unless $\underline{p}|d_-$. Suppose $\underline{p}|d_0$ and put $p^f = N\underline{p}$. Taking norms, we have that $p^{kf}|m^2 d_-$. Thus kf must be odd. Therefore $f = 1$ (p splits) and k is odd. Everything is proved except for the case $d_- = 4d_0$, d_0 odd, and $\underline{p}|2$. This requires a detailed examination of the equation $x^2 - \mu = 0$ locally at the primes dividing 2 in F. The desired result is obtained, but we omit the details.

A simple description of the whole story is

THEOREM 7. *Let $K = \underline{Q}(\sqrt{d})$ be a quadratic field of discriminant d. A necessary and sufficient condition that K have an unramified cyclic extension of degree 4 is that there exist an admissible factorization, $d = d_+ d_-$ such that d_- is a norm from $\underline{Q}(\sqrt{d_+})$ and 2 splits in $\underline{Q}(\sqrt{d_+})$ if $2|d_-$.*

Proof. If such an extension exists we have $d_- = N(m^{-1}\mu)$ and Lemma 3 says that 2 splits in $\underline{\underline{Q}}(\sqrt{d_+})$ if $2|d_-$. Now suppose the condition is fulfilled. Then $d_0 = \alpha^{1+\sigma}$ for some $\alpha \in \underline{\underline{Q}}(\sqrt{d_+})$. The prime factorization of α must have the form $(\alpha) = \underline{d}\underline{m}^{1-\sigma}$. Put $m_0 = N\underline{m}$; then $\mu = m_0\alpha$ has all the properties previously given (if this μ is not square-free, it can be made so by modifying $\underline{m}$). We may also assume that $\underline{m}$ is relatively prime to 2, for suppose $\underline{m}$ is exactly divisible by $\underline{p}^k$ where $\underline{p}|2$. Since μ is square-free, $\underline{p}$ is not principal. Therefore $\underline{p}^{1+\sigma} = (2)$ and there is an integer $\beta \in \underline{\underline{Q}}(\sqrt{d_+})$ such that $(\beta) = \underline{p}^{\sigma}\underline{q}$ where $\underline{q}$ is relatively prime to 2. We replace $\underline{m}$ by $(\beta/2)^k\underline{m}$, i.e. we change μ to $(\beta/2)^{2k}\mu$, with no loss. The equation $x^2 - \mu = 0$ does not ramify over F at any prime not dividing 2 since $(\mu) = \underline{m}^2\underline{d}$ is the square of an ideal in F. The ramification at 2 is handled this way: we have

$$\mu = (a + b\sqrt{d_+})/2 \quad , \quad a,b \in \underline{\underline{Z}} \quad \text{both even or both odd}$$

and the calculation of $N\mu$ gives

$$4m_0^2d_0 = a^2 - b^2d_+ \quad ; \quad (m_0 \text{ is odd}).$$

Let $b > 0$ be fixed such that $(\pm a,b)$ is a solution to the norm equation. We can always choose the sign of a so that $x^2 - \mu = 0$ does not ramify over F at primes dividing 2. The proof requires an analysis of cases. Case I is $d_+ \equiv 1 \bmod 4$, $d_- \equiv 1 \bmod 4$; here the condition is that μ be a quadratic residue mod 4 in $\underline{\underline{Q}}(\sqrt{d_+})$. The remaining cases are

II: $d_+ \equiv 8 \bmod 32$, $d_- \equiv 1 \bmod 4$

III: $d_+ \equiv 1 \bmod 8$, $d_- \equiv -4 \bmod 16$

III': $d_+ \equiv 1 \bmod 8$, $d_- \equiv -8 \bmod 32$.

The ramification conditions become complicated when $2|d$, and we omit the tedious details of the case analyses.

The unramified cyclic extension of degree 4 is of the form $L = \underline{\underline{Q}}(\sqrt{d_-},\sqrt{\mu})$.

Here are some examples, including all known cases where $d < 0$, $4|d$, and $\underline{\underline{H}} \simeq \underline{\underline{Z}}_4$

d	d_+	d_-	μ	$\underline{\underline{H}}$
-56	8	-7	$-1 + 2\sqrt{2}$	$\underline{\underline{Z}}_4$
-184	8	-23	$-3 + 4\sqrt{2}$	$\underline{\underline{Z}}_4$
-248	8	-31	$1 + 4\sqrt{2}$	$\underline{\underline{Z}}_8$
-568	8	-71	$-1 + 6\sqrt{2}$	$\underline{\underline{Z}}_4$
-68	17	-4	$4 + \sqrt{17}$	$\underline{\underline{Z}}_4$
-64	41	-4	$32 + 5\sqrt{41}$	$\underline{\underline{Z}}_8$
-260	65	-4	$8 + \sqrt{65}$	$\underline{\underline{Z}}_2 \times \underline{\underline{Z}}_4$
-292	73	-4	$1068 + 125\sqrt{73}$	$\underline{\underline{Z}}_4$
-356	89	-4	$500 + 53\sqrt{89}$	$\underline{\underline{Z}}_4$
-772	193	-4	$1764132 + 126985\sqrt{193}$	$\underline{\underline{Z}}_4$
-136	17	-8	$(3 + \sqrt{17})/2$	$\underline{\underline{Z}}_4$
-328	41	-8	$(19 + 3\sqrt{41})/2$	$\underline{\underline{Z}}_4$
-55	5	-11	$3 + 2\sqrt{5}$	$\underline{\underline{Z}}_4$
-95	5	-19	$-1 + 2\sqrt{5}$	$\underline{\underline{Z}}_8$
-155	5	-31	$-7 + 4\sqrt{5}$	$\underline{\underline{Z}}_4$
-39	13	-3	$(-1 + \sqrt{13})/2$	$\underline{\underline{Z}}_4$
-111	37	-3	$(-5 + \sqrt{37})/2$	$\underline{\underline{Z}}_4$
+145	29	5	$11 + 2\sqrt{29}$	$\underline{\underline{Z}}_4$

(In the real case one distinguishes d_+ from d_- by testing: here 7 remains prime in $\underline{Q}(\sqrt{145})$ and $\underline{Q}(\sqrt{5})$ but splits in $\underline{Q}(\sqrt{29})$; hence $\sqrt{29}$ is fixed under τ ; cf. §3.)

VIII COMPUTATION OF SINGULAR j-INVARIANTS

(C. Herz, Jan. 15, 1958)

Let C be an elliptic curve. By a "model" of C we shall mean here a non-singular embedding in the projective plane. The first fact to be noted is that the class of non-singular plane cubics and the class of models of elliptic curves are identical. However there is a much deeper statement available. The natural equivalence relation for algebraic curves is birational equivalence, i.e. isomorphism of the associated function fields; the natural equivalence relation for projective models is projective equivalence, i.e. two plane curves are projectively equivalent if and only if one may be transformed into the other by a collineation. Projective equivalence is a priori narrower than birational equivalence. The second main fact is that for elliptic curves the two notions coincide provided that the ground field is algebraically closed of characteristic $\neq 2$ or 3. This is substantially the same as saying that one can choose a standard cubic model for each elliptic curve which, as it turns out, depends on a single parameter. This parameter is essentially the modulus for the curve.

The general homogeneous cubic polynomial in three variables, $H(x_0, x_1, x_2)$, has nine coefficients and so the cubic curves $H = 0$ form an 8-dimensional projective space. If we subject the coordinates to a linear transformation Π, a non-singular 3×3 matrix, the coefficients undergo a transformation by the symmetrized kronecker cube of the transposed $H^{\Pi}(x) = H(\Pi x)$ of Π. The group of these transformations, obtained by taking all non-singular Π's with coefficients in the constant field, acts on the projective 8-space. The homogeneous space obtained by factorization is the variety V of all plane cubic curves. It is easy to see that H may be put in the form

$$H(x_0, x_1, x_2) \equiv ax_0x_2^2 - 4bx_1^3 + c_2x_0^2x_1 + c_3x_0^3 .$$

The remaining transformations consist of the diagonal matrices Π. Thus V is the projective 3-space (a, b, c_2, c_3) modulo the action of the group induced by the diagonal matrices Π. This reduction step simplifies the computations. What we are looking for are invariants of cubic curves, -- the first things needed are the ray-invariants, i.e. the homogeneous polynomials, g, in the coefficients of the general cubic which under the action of a matrix Π transform by being multiplied by a power of the determinant of Π. Any ray invariant, g, is of course a ray invariant in the reduced case; the converse is true by the "unitary trick" which is applicable because the general linear group modulo the diagonal subgroup is compact. Now if Π acting on $\begin{pmatrix} x_2 \\ x_1 \\ x_0 \end{pmatrix}$ is given by the matrix $\begin{pmatrix} \alpha & & 0 \\ & \beta & \\ 0 & & \gamma \end{pmatrix}$ the action on $\begin{pmatrix} a \\ b \\ c_2 \\ c_3 \end{pmatrix}$ is given by

$$\begin{pmatrix} \alpha^2\gamma & & & 0 \\ & \beta^3 & & \\ & & \beta\gamma^2 & \\ 0 & & & \gamma^3 \end{pmatrix} .$$

Since this is diagonal, a basis for the ray invariants of degree m is formed by the monomials $g = a^u b^v c_2^x c_3^y$ where $u + v + x + y = m$. Since we demand $g^{\Pi} = (\det \Pi)^m g$ we have the equations $2u = m$, $3v + x = m$, $u + 2x + 3y = m$. The general solution is $v = u - z$, $x = -u + 3z$, $y = u - 2z$ where z is another integer and $3z \geq u \geq 2z$. Taking $z = 1$ there are two possibilities:

$u = 2$ corresponding to the monomial $g_2 = a^2bc_2$ and $u = 3$ corresponding to $g_3 = a^3b^2c_3$. It follows that all other ray invariants are polynomials in g_2 and g_3.

$\Delta = g_2^3 - 27g_3^2 = a^6b^3\ (c_2^3 - 27bc_3^2)$ is another ray invariant. It has the virtue that $\Delta = 0$ is a necessary and sufficient condition for the cubic curve to be singular, i.e. either have a linear factor or a double point. $j = 2^6 \cdot 3^3\ g_2^3 \Delta^{-1}$ is defined for all non-singular cubics. If V_o is the sub-variety of V defined by $\Delta = 0$ then it is easy to see that $V - V_o$ is an affine line and j is a uniformizing parameter. $V - V_o$ is the variety of moduli of elliptic curves.

Δ has another virtue. Suppose $H(x_o, x_1, x_2) = 0$ is an elliptic curve, C. Then

$$du = \frac{2 \begin{vmatrix} \lambda_o & x_o & dx_o \\ \lambda_1 & x_1 & dx_1 \\ \lambda_2 & x_2 & dx_2 \end{vmatrix}}{\Sigma_{j=o}^{2} \lambda_j \partial H / \partial x_j}$$

is a differential of the first kind on C, and is independent of the choice of the point $(\lambda_2, \lambda_1, \lambda_o)$ in the projective plane. Let us examine the effect of a change of coordinates on du. Suppose $x = \Pi x^*$. Then the new equation is $H^*(x^*) = H(\Pi x^*) = H^{\Pi}(x^*)$. In the new system

$$du^* = \frac{2 \begin{vmatrix} \lambda_o^* & x_o^* & dx_o^* \\ \lambda_1^* & x_1^* & dx_1^* \\ \lambda_2^* & x_2^* & dx_2^* \end{vmatrix}}{\Sigma_{j=o}^{2} \lambda_j^* \partial H^* / \partial x_j^*} .$$

Of course $du^* = \mu du$ where μ is a constant multiplier since differentials of the first kind differ only by a constant factor. However since λ is arbitrary we can put $\lambda = \Pi \lambda^*$ and it becomes evident that $du = (\det \Pi) du^*$. Hence $\mu = \det \Pi$ and $\mu^{12} = \Delta^*/\Delta$, -- for this reason expressions of the form Δ^*/Δ are occasionally called multipliers. Δdu^{12} is an invariant of the curve. Could we choose $\Delta^{1/12}$ invariantly, $du_C = \Delta^{1/12} du$ would be a differential of the first kind canonically attached to the curve.

The group operation on an elliptic curve has a simple geometric interpretation. Take a model and let $(\mathfrak{u})$ and $(\mathfrak{v})$ be two points on it. These points determine a line which intersects the cubic in exactly one other point which we write as $(-\mathfrak{u}-\mathfrak{v})$. The curve has a unique point of inflection $(\mathcal{O})$. It is easy to check that the operations defined give rise to an additive abelian group with identity element $(\mathcal{O})$. Since the group operation is defined by lines it is clearly invariant under collineations.

There are exactly three points on C, other than $(\mathcal{O})$, whose tangents pass through $(\mathcal{O})$. If $(\mathfrak{n})$ is such a point, $(2\mathfrak{n}) = (\mathcal{O})$, according to the description of the addition operation, and the converse is true. Thus if $(\mathfrak{n}_1)$ and $(\mathfrak{n}_2)$ are two such points, $(\mathfrak{n}_3) = (-\mathfrak{n}_1 - \mathfrak{n}_2)$ is a third. These three points are collinear; call the line they determine L_C. Projection from $(\mathcal{O})$ gives a two-fold covering of C onto L_C; two points on C go into the same point of L_C if and only if they are inverses of one another. (The projection is not defined at $(\mathcal{O})$ so we extend by continuity, -- $(\mathcal{O})$ is projected along the inflection tangent.) Let $\mathfrak{n}_o$ be the projection of $\mathcal{O}$ on L_C. The cross ratio $(\mathfrak{n}_o\mathfrak{n}_1\mathfrak{n}_2\mathfrak{n}_3) = k^2$ is invariant under projective transformations; however the three points $(\mathfrak{n}_1)$, $(\mathfrak{n}_2)$, $(\mathfrak{n}_3)$ may be permuted at will so

that k^2 is an invariant of C only up to the action of the symmetric group on three letters. If we write $k'^2 = 1 - k^2$, the full invariant is
$j = 2^8 \frac{(1 - k^2k'^2)^3}{k^4k'^4}$.

Suppose C^* is an elliptic curve which forms an m-fold covering of C. Since the covering map must be a homomorphism of the group structures with suitable base points, inverses are mapped into inverses. Thus if we have a model of C^* with inflection point $(\mathcal{O}^*)$ mapping onto a model of C with inflection point $(\mathcal{O})$, there is a well-defined map of L_{C^*} onto L_C which is a m-fold covering of one projective line by another. It will be convenient to take the models in the projective plane with $(\mathcal{O})$ at $x_o = x_1 = 0$, $x_2 = 1$, inflection tangent $x_o = 0$, and the line L_C as $x_2 = 0$. The covering map of L_{C^*} onto L_C then has the form

$$x_o = Q(x_o^*, x_1^*)x_o^* , \qquad x_1 = P(x_o^*, x_1^*)$$

where Q is a homogeneous polynomial of degree m - 1 and P is one of degree m. The covering map of C^* onto C may then be described at all but a finite number of points by the additional transformation $x_2 = R(x_o^*, x_1^*)x_2^*$ where R is a homogeneous rational function of degree n - 1. We shall now find the explicit form of R. The equations of C and C^* are $x_o x_2^2 - T(x_o, x_1) = 0$ and $x_o^* x_2^{*2} - T^*(x_o^*, x_1^*) = 0$ respectively where T and T^* are homogeneous cubics with the coefficient of x_1^3 different from 0. The differentials of the first kind are

$$du = \frac{\begin{vmatrix} x_o & dx_o \\ x_1 & dx_1 \end{vmatrix}}{x_o x_2} \quad \text{and} \quad du^* = \frac{\begin{vmatrix} x_o^* & dx_o^* \\ x_1^* & dx_1^* \end{vmatrix}}{x_o^* x_2^*} .$$

Since x_o, x_1 are homogeneous polynomials of degree m in x_o^*, x_1^*,

$$\begin{vmatrix} x_o & dx_o \\ x_1 & dx_1 \end{vmatrix} = \frac{1}{m} \left| \frac{\partial(x_o, x_1)}{\partial(x_o^*, x_1^*)} \right| \cdot \begin{vmatrix} x_o^* & dx_o^* \\ x_1^* & dx_1^* \end{vmatrix} .$$

Hence $du = \frac{1}{m} \left| \frac{\partial(x_o, x_1)}{\partial(x_o^*, x_1^*)} \right| \frac{du^*}{QR}$, but we must also have $du^* = \mu du$ where μ is a constant multiplier. This tells us that $R = \frac{\mu}{m} Q^{-1} \left| \frac{\partial(x_o, x_1)}{\partial(x_o^*, x_1^*)} \right|$ $= \frac{\mu}{m} \left\{ \frac{\partial P}{\partial x_1^*} + Q^{-1} x_o^* \left| \frac{\partial(Q, P)}{\partial(x_o^*, x_1^*)} \right| \right\}$. Substituting in the equation for the curves we find that

$$\frac{\mu^2}{m^2} \left\{ Q\left(\frac{\partial P}{\partial x_1^*}\right)^2 + 2x_o^* \left| \frac{\partial(Q,P)}{\partial(x_o^*,x_1^*)} \right| + Q^{-1} x_o^{*2} \left| \frac{\partial(Q,P)}{\partial(x_o^*, x_1^*)} \right|^2 \right\} \cdot T^*(x_o^*, x_1^*)$$

$$= T(x_o^* Q, P).$$

There can be no denominator arising from the Q^{-1} term on the left. Hence the only possible linear factors of Q are x_o^* and the factors of T^*; all other factors of Q must be at least quadratic.

What we have just described is the Jacobi transformation principle. It leads to the algebraic computation of the singular class invariants j. Some explicit computations are performed in Weber: Lehrbuch der Algebra, Vol. III, sections 8-10. The results, of course, coincide with the invariant equations for j obtained by analytic means. These equations will hold in any of the fields we have considered, i.e. the analytic method gives results valid over rather general ground fields. The difficulty in making computations from the class equations is due to the fact that the coefficients in the q-expansion of j are intractably large:
$j = q^{-1} + 744 + 196,884q + 21,493,760q^2 + \dots$. It is in fact easier to

compute using k^2. However we shall now proceed to the analytic case and examine the class equation by investigating the subgroups of finite index in the modular group.

Let Γ denote the full modular group and H a (not necessarily normal) subgroup of index ψ. We shall write $\Gamma = \Sigma_{a=1}^{\psi} HV_a$. Let $\mathfrak{D}$ be the standard fundamental domain for the modular group; $\mathfrak{D}$ has a simplicial decomposition with 2 faces, 3 edges, and 3 vertices, namely $\tau = i\infty$, i, and ρ. S and T stand for the transformations S: $\tau \longrightarrow \tau + 1$, T: $\tau \longrightarrow - 1/\tau$. $i\infty$ is a fixed point of S and all its powers, i is a fixed point of T ($T^2 = I$), ρ is a fixed point of $S^{-1}T$ ($(S^{-1}T)^3 = I$). Γ is generated by S and T and the only relations are consequences of those given above.

A fundamental domain, $\mathfrak{F}$, for the subgroup H is defined as follows. At each interior point $\tau \in \mathfrak{D}$ we assign ψ points $V_1\tau, \ldots V_\psi\tau$ in $\mathfrak{F}$. These points are inequivalent modulo H for if $V_a\tau = UV_b\tau$ then τ is a fixed point of $V_a^{-1}UV_b$. This can occur for $\tau \in \mathfrak{D}$, $\tau \neq i\infty$, i, ρ only if $V_a^{-1}UV_b = I$. Hence if $U \in H$, $V_a = UV_b$ contrary to the assumption that for $a \neq b$, V_a and V_b represent different cosets. Thus $\mathfrak{F}$ has 2ψ faces and 3ψ edges.

The vertices require special treatment. First we take $\tau = i\infty$. Let us define an equivalence relation $V_a \underset{\infty}{\sim} V_b$ if and only if $V_aS^kV_b^{-1} \in H$ for some integer k. This provides a grouping of the cosets of Γ mod H into equivalence classes $K_1, \ldots K_d$ with K_c/H of order n_c. For each class K_c there is one vertex of $\mathfrak{F}$ above $i\infty$ in $\mathfrak{D}$ at which n_c sheets come together.

A similar but simpler analysis holds at $\tau = i$ and $\tau = \rho$. V_a and $V_a T$ represent two different sheets of $\mathfrak{F}$ unless $V_a T V_a^{-1} \varepsilon H$; let ε_i be the number of solutions of this equation. Then there are $\frac{1}{2}(\psi - \varepsilon_i)$ vertices of $\mathfrak{F}$ above i in $\mathfrak{D}$ at which 2 sheets come together and ε_i unbranched vertices. Likewise V_a, $V_a(S^{-1}T)$, and $V_a(S^{-1}T)^2$ represent three distinct sheets of $\mathfrak{F}$ unless $V_a S^{-1} T V_a^{-1} \varepsilon H$ in which case all three coincide; let ε_ρ be the number of solutions of the last equation. There are $\frac{1}{3}(\psi - \varepsilon_\rho)$ vertices of $\mathfrak{F}$ above ρ in $\mathfrak{D}$ at which 3 sheets come together and ε_ρ unbranched vertices.

The total number of vertices of $\mathfrak{F}$ is $d + \frac{1}{2}(\psi + \varepsilon_i) + \frac{1}{3}(\psi + 2\varepsilon_\rho)$. Hence, in particular, $\mathfrak{F}$ is a sphere if and only if $2\psi - 3\psi + d + \frac{1}{2}(\psi + \varepsilon_i) + \frac{1}{3}(\psi + 2\varepsilon_\rho) = 2$, i.e. $\psi = 6d + 3\varepsilon_i + 4\varepsilon_\rho - 12$.

The most important subgroups for our purposes are the transformation groups H_m and the congruence groups Γ_m. The transformation groups arise as follows. Let M be the matrix $\begin{pmatrix} m & 0 \\ 0 & 1 \end{pmatrix}$. $U \varepsilon H_m$ if $U \varepsilon \Gamma$ and $MUM^{-1} \varepsilon \Gamma$ which is equivalent to saying that U has the matrix representation $U = \begin{pmatrix} \alpha & \beta \\ \gamma & \delta \end{pmatrix}$ with $\alpha, \beta, \gamma, \delta$ integers, $\alpha\delta - \beta\gamma = 1$ and $\gamma \equiv 0 \bmod m$. Writing $\Gamma = \Sigma_{a=1}^{\psi(m)} H_m V_a$ the matrices MV_a can be chosen to run through the set $\begin{pmatrix} \alpha & \beta \\ 0 & \delta \end{pmatrix}$ where $\alpha \geq 1$, $0 \leq \beta < \delta$, $\alpha\delta = m$ and $(\alpha, \beta, \delta) = 1$. We shall write $V_{\alpha,\beta,\delta}$ for V_a. The invariant polynomial is $J(t, j) = \prod_{\alpha,\beta,\delta} (t - j^{MV_{\alpha,\beta,\delta}})$. j^M defined by $j^M(\tau) = j(m\tau)$ is invariant under the action of H_m. Thus j^M is a meromorphic function on the Riemann surface $\mathfrak{F}_m$ corresponding to H_m. The Galois group of the invariant equation $J(t, j) = 0$ over the field $Q(j)$ is Γ/Γ_m where Γ_m is the largest subgroup of H_m which is normal in Γ. Γ_m is represented by the matrices $\begin{pmatrix} \alpha & \beta \\ \gamma & \delta \end{pmatrix} \equiv \begin{pmatrix} 1 & 0 \\ 0 & 1 \end{pmatrix} \bmod m$ with $\alpha, \beta, \gamma, \delta$

integers and $\alpha\delta - \beta\gamma = 1$. If m is a prime Γ/Γ_m is isomorphic to $SL_2(Z_m)$ modulo its center; this group is simple for $m \geq 5$.

The computations involving singular invariants j are easy to carry out when the Riemann surface $\mathcal{F}_m$ corresponding to H_m is a sphere. Here $\psi = \psi(m) = m \prod_{p|m} (1 + \frac{1}{p})$. We have to calculate d, ε_i, and ε_ρ. The equivalence relation at $i\infty$, $V_{\alpha,\beta,\delta} \sim V_{\alpha',\beta',\delta'}$ may be formulated as $\begin{pmatrix}\alpha & \beta\\ 0 & \delta\end{pmatrix}\begin{pmatrix}1 & k\\ 0 & 1\end{pmatrix} = U\begin{pmatrix}\alpha' & \beta'\\ 0 & \delta'\end{pmatrix}$ where $U \in \Gamma$. The matrix on the left is $\begin{pmatrix}\alpha & \beta+k\alpha\\ 0 & \delta\end{pmatrix}$ and the equivalence holds only if $\alpha' = \alpha$, $\delta' = \delta$ and $\beta' \equiv \beta + k\alpha$ mod δ. For simplicity assume m is square-free. Then $(\alpha, \delta) = 1$ and β' can run through all the values $0, 1, \ldots \delta-1$ for any given β. Hence for each $\delta | m$ we have an equivalence class K_δ containing δ cosets. The number d is d(m), the number of divisors of m.

The equation $V_{\alpha,\beta,\delta}\, T V^{-1}_{\alpha,\beta,\delta} \in H_m$ is equivalent to $\delta = m$ and $\beta^2 + 1 \equiv 0$ mod m. The analogous equation at the vertices above $\tau = \rho$ is equivalent to $\delta = m$ and $\beta^2 - \beta + 1 \equiv 0$ mod m. For simplicity, suppose from now on that m is a prime > 3. Then $\psi(m) = m+1$, $d = 2$, $\varepsilon_i = 2$ or 0 according to whether -1 is a quadratic residue mod m or not, $\varepsilon_\rho = 2$ or 0 according to whether -3 is a quadratic residue mod m or not. The condition for $\mathcal{F}_m$ to be a sphere is $m+1 = 3\varepsilon_i + 4\varepsilon_\rho$. This holds for m = 5, 7, 13 and no other primes > 3.

We shall now carry out an extended analysis for the case m = 5. On $\mathcal{F}_5$ there is one vertex above $\tau = i\infty$ at which there is only one sheet and one vertex at which 5 sheets come together. Above $\tau = i$ there are two vertices at which 2 sheets come together and two unbranched vertices. Above $\tau = \rho$ there are two vertices at which three sheets come together.

We know in advance that $\frac{\Delta(\frac{1}{5}\omega_1, \omega_2)}{\Delta(\omega_1, \omega_2)}$ is a meromorphic function on $\mathfrak{F}_m$ since it is invariant under H_m. The only possible zeros and poles occur at the vertices lying over $\tau = i\infty$. Let q be the local uniformizing parameter $e^{2\pi i \omega_2/\omega_1}$ at the unbranched vertex over $\tau = i\infty$. Here

$$\frac{\Delta(\frac{1}{5}\omega_1, \omega_2)}{\Delta(\omega_1, \omega_2)} = 5^{12} q^4 \prod_{5\nmid n} (1-q^n)^{-24} .$$

The function has a four-fold zero; since it has only one pole, the pole is also 4-fold and this function is the fourth power of a function f with only one single pole. Since $\mathfrak{F}_5$ is a sphere, f is a global uniformizing parameter. Now consider the function fj. f has a six-fold pole at the branched vertex above $\tau = i\infty$. At each of the two vertices above $\tau = \rho$ it has a three-fold zero. Hence $fj = v^3$ where v is a meromorphic function on $\mathfrak{F}_5$. Since v has a double pole at the pole of f, v is in fact a quadratic polynomial in f. Let

$$q' = e^{-2\pi i \frac{\omega_1}{5\omega_2}}$$

be the local uniformizing parameter at the branched vertex above $\tau = i\infty$. At this point $j = q'^{-5} + E$ where E is a power series in non-negative powers of q'. f is defined by

$$f = \left\{ \frac{\Delta(-\omega_2, \frac{1}{5}\omega_1)}{\Delta(-\omega_2, \omega_1)} \right\}^{1/4} = q'^{-1} \prod_{5\nmid n} (1-q'^n)^6$$

so that $f = q'^{-1} + \dots$, $f^2 = q'^{-2} - 12q'^{-1} + \dots$, and $v = q'^{-2} - 2q'^{-1} + \dots$. Hence $v - f^2 - 10f$ has no poles; it is thus a constant. At the other vertex $j = q^{-1} + \dots$, $f = 5^3 q + \dots$ so $v = 5 + \dots$ and the constant term is 5. We have computed $v = f^2 + 10f + 5$, $j = \frac{(f^2 + 10f + 5)^3}{f}$.

We now make use of a very important property of the transformation groups: namely $(TM)H_m(TM)^{-1} = H_m$. Thus $f^{(TM)}$ is a meromorphic function on $\mathfrak{J}_m$ if f is. In the case $m = 5$ let us write $\tilde{f} = f^{TM}$, i.e. $\tilde{f}(\tau) = f(-\frac{1}{5\tau})$. The q-expansion of $\tilde{f}$ is the same as the q'-expansion of f, i.e. $\tilde{f} = q^{-1} + \ldots$. Since $\tilde{\tilde{f}} = f$, $\tilde{f}$ has only one pole, and we conclude that $\tilde{f} = 5^3 f^{-1}$. For $\tau = \sqrt{-5}/5$, $-\frac{1}{5\tau} = \tau$ so $f = \tilde{f}$ and hence $f(\sqrt{-5}/5) = \pm 5^{3/2}$. With either choice of sign $f^2 + 10f + 5 > 0$; hence the sign is the same as that of $j(\sqrt{-5}/5)$, but j is positive on the imaginary axis. Accordingly $f(\sqrt{-5}/5) = + 5^{3/2}$. $(5, \sqrt{-5})$ is an ideal basis in the field $\mathbb{Q}(\sqrt{-5})$ representing the principal class. From the expression of j in terms of f one has $j = 2^3 \cdot 5 \cdot (25 + 13\sqrt{5})^3$ for the singular invariant of the principal class.